ADVANCED ELECTRIC DRIVES

ADVANCED ELECTRIC DRIVES

Analysis, Control, and Modeling Using MATLAB/Simulink®

Ned Mohan

WILEY

Published by John Wiley & Sons, Inc., Hoboken, New Jersey.
Published simultaneously in Canada.

MATLAB and Simulink are registered trademarks of The MathWorks, Inc. See www.mathworks.com/trademarks for a list of additional trademarks. **The MathWorks Publisher Logo identifies books that contain MATLAB® content. Used with permission. The MathWorks does not warrant the accuracy of the text or exercises in this book or in the software downloadable from** http://www.wiley.com/WileyCDA/WileyTitle/productCd-047064477X.html **and** http://www.mathworks.com/matlabcentral/fileexchange/?term=authorid%3A80973. **The book's or downloadable software's use or discussion of MATLAB® software or related products does not constitute endorsement or sponsorship by The MathWorks of a particular use of the MATLAB® software or related products.**

For MATLAB® and Simulink® product information, or information on other related products, please contact:

The MathWorks, Inc.
3 Apple Hill Drive
Natick, MA 01760-2098 USA
Tel 508-647-7000
Fax: 508-647-7001
E-mail: info@mathworks.com
Web: www.mathworks.com

For general information on our other products and services or for technical support, please contact our Customer Care Department within the United States at (800) 762-2974, outside the United States at (317) 572-3993 or fax (317) 572-4002.

Wiley also publishes its books in a variety of electronic formats. Some content that appears in print may not be available in electronic formats. For more information about Wiley products, visit our web site at www.wiley.com.

Library of Congress Cataloging-in-Publication Data:

Mohan, Ned.
  Advanced electric drives : analysis, control, and modeling using MATLAB/Simulink® / Ned Mohan.
       pages cm
    Includes index.
   ISBN 978-1-118-48548-4 (hardback)
1. Electric driving–Computer simulation.   2. Electric motors–Mathematical models.   3. MATLAB.   4. SIMULINK.   I. Title.
   TK4058.M5783 2014
   621.460285'53–dc23
                        2014005496

Printed in the United States of America.

10 9 8 7 6 5 4 3 2 1

CONTENTS

3 Dynamic Analysis of Induction Machines in Terms of *dq* Windings **28**

PREFACE

When I wrote the first version of this textbook in 2001, my opening paragraph was as follows:

> Why write a textbook for a course that has pretty much disappeared from the curriculum at many universities? The only possible answer is in hopes of reviving it (as we have been able to do at the University of Minnesota) because of enormous future opportunities that await us including biomedical applications such as heart pumps, harnessing of renewable energy resources such as wind, factory automation using robotics, and clean transportation in the form of hybrid-electric vehicles.

Here we are, more than a decade later, and unfortunately the situation is no different. It is hoped that the conditions would have changed when the time comes for the next revision of this book in a few years from now.

This textbook follows the treatment of electric machines and drives in my earlier textbook, *Electric Machines and Drives: A First Course*, published by Wiley (http://www.wiley.com/college/mohan).

My attempt in this book is to present the analysis, control, and modeling of electric machines as simply and concisely as possible, such that it can easily be covered in one semester graduate-level course. To do so, I have chosen a two-step approach: first, provide a "physical" picture without resorting to mathematical transformations for easy visualization, and then confirm this physics-based analysis mathematically.

The "physical" picture mentioned above needs elaboration. Most research literature and textbooks in this field treat dq-axis transformation of a-b-c phase quantities on a purely mathematical basis, without relating this transformation to a set of windings, albeit hypothetical, that can be visualized. That is, we visualize a set of hypothetical dq windings along an orthogonal set of axes and then relate their currents and voltages to the a-b-c phase quantities. This discussion follows seamlessly from the treatment of space vectors and the equivalent winding representations in steady state in the previous course and the textbook mentioned earlier.

For discussion of all topics in this course, computer simulations are a necessity. For this purpose, I have chosen MATLAB/Simulink® for the following reasons: a student-version that is more than sufficient for our purposes is available at a very reasonable price, and it takes extremely short time to become proficient in its use. Moreover, this same software simplifies the development of a real-time controller of drives in the hardware laboratory for student experimentation—such a laboratory using 42-V machines is developed using digital control and promoted by the University of Minnesota. The MATLAB and Simulink files used in examples are included on the accompanying website to this textbook: www.wiley.com/go/advancedelectricdrives.

As a final note, this textbook is not intended to cover power electronics and control theory. Rather, the purpose of this book is to analyze electric machines in a way that can be interfaced to well-known power electronic converters and controlled using any control scheme, the simplest being proportional-integral control, which is used in this textbook.

NED MOHAN
University of Minnesota

NOTATION

1. Variables that are functions of time $\qquad$ v, i, λ
2. Peak values (of time-varying variables) $\qquad$ $\hat{V}, \hat{I}, \hat{\lambda}$
3. Phasors $\qquad$ $\bar{V} = \hat{V}\angle\theta_v, \bar{I} = \hat{I}\angle\theta_i$
4. Space vectors $\qquad$ $\vec{H}(t), \vec{B}(t), \vec{F}(t), \vec{v}(t) = \hat{V}e^{j\theta}, \vec{i}(t) = \hat{I}e^{j\theta}, \vec{\lambda}(t) = \hat{\lambda}e^{j\theta}$
   For space vectors, the exponential notion is used where,

$$e^{j\theta} = 1\angle\theta = \cos\theta + j\sin\theta e^{j\theta} = 1\angle\theta = \cos\theta + j\sin\theta.$$

Note that both phasors and space vectors, two distinct quantities, have their peak values indicated by "^."

SUBSCRIPTS

Stator phases	a, b, c
Rotor phases	A, B, C
dq windings	d, q
Stator	s
Rotor	r
Magnetizing	m
Mechanical	m (as in θ_m or ω_m)
Mechanical	$mech$ (as in θ_{mech} or ω_{mech})
Leakage	ℓ

SUPERSCRIPTS

Denotes the axis used as reference for defining a space vector (lack of superscript implies that the d-axis is used as the reference).

* $\qquad$ Reference Value

SYMBOLS

p Number of poles ($p \geq 2$, even number)

θ All angles, such as θ_m and the axes orientation (for example, $e^{j2\pi/3}$), are in electrical radians (electrical radians equal $p/2$ times the mechanical radians).

ω All speeds, such as ω_{syn}, ω_d, ω_{dA}, ω_m, and ω_{slip} (except for ω_{mech}), are in electrical radians per second.

ω_{mech} The rotor speed is in actual (mechanical) radians per second: $\omega_{mech} = (2/p)\omega_m$.

θ_{mech} The rotor angle is in actual (mechanical) radians per second: $\theta_{mech} = (2/p)\theta_m$.

fl Flux linkages are represented by fl in MATLAB and Simulink examples.

INDUCTION MOTOR PARAMETERS USED INTERCHANGEABLY

$$R'_r = R_r$$
$$L'_{\ell r} = L_{\ell r}$$

1 Applications: Speed and Torque Control

There are many electromechanical systems where it is important to precisely control their torque, speed, and position. Many of these, such as elevators in high-rise buildings, we use on daily basis. Many others operate behind the scene, such as mechanical robots in automated factories, which are crucial for industrial competitiveness. Even in general-purpose applications of adjustable-speed drives, such as pumps and compressors systems, it is possible to control adjustable-speed drives in a way to increase their energy efficiency. Advanced electric drives are also needed in wind-electric systems to generate electricity at variable speed, as described in Appendix 1-A in the accompanying website. Hybrid-electric and electric vehicles represent an important application of advanced electric drives in the immediate future. In most of these applications, increasing efficiency requires producing maximum torque per ampere, as will be explained in this book. It also requires controlling the electromagnetic toque, as quickly and as precisely as possible, illustrated in Fig. 1-1, where the load torque T_{Load} may take a step-jump in time, in response to which the electromagnetic torque produced by the machine T_{em} must also take a step-jump if the speed ω_m of the load is to remain constant.

1-1 HISTORY

In the past, many applications requiring precise motion control utilized dc motor drives. With the availability of fast signal processing capability, the role of dc motor drives is being replaced by ac motor drives. The

*Advanced Electric Drives: Analysis, Control, and Modeling Using
MATLAB/Simulink®*, First Edition. Ned Mohan.
© 2014 John Wiley & Sons, Inc. Published 2014 by John Wiley & Sons, Inc.

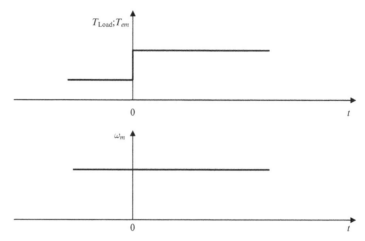

Fig. 1-1   Need for controlling the electromagnetic torque T_{em}.

use of dc motor drives in precise motion control has already been discussed in the introductory course using the textbook [1] especially designed for this purpose. Hence, our emphasis in this book for an advanced course (designed at a graduate level but that can be easily followed by undergraduates) will be entirely on ac motor drives.

1-2   BACKGROUND

In the introductory course [1], we discussed electric drives in an integrative manner where the theory of electric machines was discussed using space vectors to represent sinusoidal field distribution in the air gap. This discussion included a brief introduction to power-processing units (PPUs) and feedback control systems. In this course, we build upon that discussion and discover that it is possible to understand advanced control of electric drives on a "physical" basis, which allows us to visualize the control process rather than leaving it shrouded in mathematical mystery.

1-3   TYPES OF AC DRIVES DISCUSSED AND THE SIMULATION SOFTWARE

In this textbook, we will discuss all types of ac drives and their control in common use today. These include induction-motor drives, permanent-magnet ac drives and switched-reluctance drives. We will also discuss encoder-less operation of induction-motor drives.

A simulation-based study is essential for discussing advanced electric drive systems. After a careful review of the available software, the author considers MATLAB/Simulink® to be ideal for this purpose—a student version that is more than sufficient for our purposes is available [2] at a very reasonable price, and it takes extremely short time to become proficient in its use. Moreover, the same software simplifies the development of a real-time controller of drives in the hardware laboratory for student experimentation—such a laboratory, using 42-V machines is being developed at the University of Minnesota using digital control.

1-4 STRUCTURE OF THIS TEXTBOOK

Chapter 1 has introduced advanced electric drives. Chapter 2, Chapter 3, Chapter 4, Chapter 5, Chapter 6, Chapter 7 and Chapter 9 deal with induction-motor drives.

Chapter 8 deals with the synthesis of stator voltage vector, supplied by the inverter of the PPU, using a digital signal processor.

The permanent-magnet ac drives (ac servo drives) are discussed in Chapter 10 and the switched-reluctance motor drives are discussed in Chapter 11.

A "test" motor is selected for discussing the design of controllers and for obtaining the performance by means of simulation examples for which the specifications are provided in the next section. In all chapters dealing with induction motor drives, the "test" induction motor used is described in the following section. The "test" motor for a permanent-magnet ac drive is described in Chapter 10.

1-5 "TEST" INDUCTION MOTOR

For analyzing the performance of various control procedures, we will select a 1.5-MW induction machine as a "test" machine, for which the specifications are as follows:

Power: 1.5 MW
Voltage: 690 V (L-L, rms)

Frequency: 60 Hz
Phases: 3
Number of Poles: 6
Full-Load Slip   1%
Moment of Inertia   70 kg·m²
Per-Phase Circuit Parameters:

$R_s = 0.002 \, \Omega$

$R_r = 0.0015 \, \Omega$

$X_{\ell s} = 0.05 \, \Omega$

$X_{\ell r} = 0.047 \, \Omega$

$X_m = 0.86 \, \Omega.$

1-6   SUMMARY

This chapter describes the application of advanced ac motor drives and the background needed to undertake this study. The structure of this textbook is described in terms of chapters that cover all types of ac motor drives in common use. An absolute need for using a computer simulation program in a course like this is pointed out, and a case is made for using a general-purpose software, MATLAB/Simulink®. Finally, the parameters for a "test" induction machine are described — this machine is used in induction machine related chapters for analysis and simulation purposes.

REFERENCES

1. N. Mohan, *Electric Machines and Drives: A First Course*, Wiley, Hoboken, NJ, 2011. http://www.wiley.com/college/mohan.
2. http://www.mathworks.com.

PROBLEMS

1-1   Read the report "Adaptive Torque Control of Variable Speed Wind Turbines" by Kathryn E. Johnson, National Renewable

Energy Laboratory (http://www.nrel.gov). Upon reading section 2.1, describe the Standard Region 2 Control and describe how it works in your own words.

1-2 Read the report "Final Report on Assessment of Motor Technologies for Traction Drives of Hybrid and Electric Vehicles" (http://info.ornl.gov/sites/publications/files/pub28840.pdf) and answer the following questions for HEV/EV applications:

   (a) What are the types of machines considered?
   (b) What type of motor is the most popular choice?
   (c) What are the alternatives if NdFeB magnets are not available?
   (d) What are the advantages and disadvantages of SR motors?

1-3 Read the report "Evaluation of the 2010 Toyota Prius Hybrid Synergy Drive System" (http://info.ornl.gov/sites/publications/files/Pub26762.pdf) and answer the following questions:

   (a) What are ECVT, PCU, and ICE?
   (b) What type of motor is used in this application?

2 Induction Machine Equations in Phase Quantities: Assisted by Space Vectors

2-1 INTRODUCTION

In ac machines, the stator windings are intended to have a sinusoidally distributed conductor density in order to produce a sinusoidally distributed field distribution in the air gap. In the squirrel-cage rotor of induction machines, the bar density is uniform. Yet the currents in the rotor bars produce a magnetomotive force (mmf) that is sinusoidally distributed. Therefore, it is possible to replace the squirrel-cage with an equivalent wound rotor with three sinusoidally distributed windings.

In this chapter, we will briefly review the sinusoidally distributed windings and then calculate their inductances for developing equations for induction machines in phase (*a-b-c*) quantities. The development of these equations is assisted by space vectors, which are briefly reviewed. The analysis in this chapter establishes the framework for the *dq* winding-based analysis of induction machines under dynamic conditions carried out in the next chapter.

2-2 SINUSOIDALLY DISTRIBUTED STATOR WINDINGS

In the following analysis, we will also assume that the magnetic material in the stator and the rotor is operated in its linear region and has an infinite permeability.

Advanced Electric Drives: Analysis, Control, and Modeling Using MATLAB/Simulink®, First Edition. Ned Mohan.
© 2014 John Wiley & Sons, Inc. Published 2014 by John Wiley & Sons, Inc.

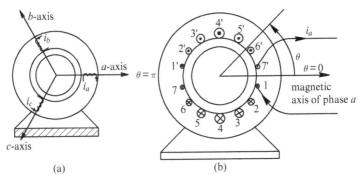

Fig. 2-1 Stator windings.

In ac machines of Fig. 2-1a, windings for each phase ideally should produce a sinusoidally distributed radial field (F, H, and B) in the air gap. Theoretically, this requires a sinusoidally distributed winding in each phase. If each phase winding has a total of N_s turns (i.e., $2N_s$ conductors), the conductor density $n_s(\theta)$ in phase-a of Fig. 2-1b can be defined as

$$n_s(\theta) = \frac{N_s}{2}\sin\theta, \quad 0 \le \theta \le \pi. \tag{2-1}$$

The angle θ is measured in the counter-clockwise direction with respect to the phase-a magnetic axis. Rather than restricting the conductor density expression to a region $0 < \theta < \pi$, we can interpret the negative of the conductor density in the region $\pi < \theta < 2\pi$ in Eq. (2-1) as being associated with carrying the current in the opposite direction, as indicated in Fig. 2-1b.

In a multi-pole machine (with $p > 2$), the peak conductor density remains $N_s/2$, as in Eq. (2-1) for a two-pole machine, but the angle θ is expressed in electrical radians. Therefore, we will always express angles in all equations throughout this book by θ in electrical radians, thus making the expressions for field distributions and space vectors applicable to two-pole as well as multi-pole machines. For further discussion on this, please refer to example 9-2 in Reference [1].

The current i_a through this sinusoidally distributed winding results in the air gap a magnetic field (mmf, flux density, and field intensity) that is co-sinusoidally distributed with respect to the position θ away from the magnetic axis of the phase

$$H_a(\theta) = \frac{N_s}{p\ell_g} i_a \cos\theta \qquad (2\text{-}2)$$

$$B_a(\theta) = \mu_o H_a(\theta) = \left(\frac{\mu_o N_s}{p\ell_g}\right) i_a \cos\theta \qquad (2\text{-}3)$$

and

$$F_a(\theta) = \ell_g H_a(\theta) = \frac{N_s}{p} i_a \cos\theta. \qquad (2\text{-}4)$$

The radial field distribution in the air gap peaks along the phase-*a* magnetic axis, and at any instant of time, the amplitude is linearly proportional to the value of i_a at that time. Notice that regardless of the positive or the negative current in phase-*a*, the flux-density distribution produced by it in the air gap always has its peak (positive or negative) along the phase-*a* magnetic axis.

2-2-1 Three-Phase, Sinusoidally Distributed Stator Windings

In the previous section, we focused only on phase-*a*, which has its magnetic axis along $\theta = 0°$. There are two more identical sinusoidally distributed windings for phases *b* and *c*, with magnetic axes along $\theta = 120°$ and $\theta = 240°$, respectively, as represented in Fig. 2-2a. These three

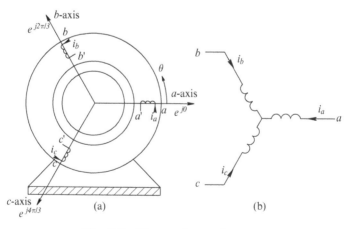

Fig. 2-2 Three-phase windings.

windings are generally connected in a wye-arrangement by joining terminals a', b', and c' together, as shown in Fig. 2-2b. A positive current into a winding terminal is assumed to produce flux in the radially outward direction. Field distributions in the air gap due to currents i_b and i_c are identical in sinusoidal shape to those due to i_a, but they peak along their respective phase-b and phase-c magnetic axes.

2-3 STATOR INDUCTANCES (ROTOR OPEN-CIRCUITED)

The stator windings are assumed to be wye-connected as shown in Fig. 2-2b where the neutral is not accessible. Therefore, at any time

$$i_a(t) + i_b(t) + i_c(t) = 0. \qquad (2\text{-}5)$$

For defining stator-winding inductances, we will assume that the rotor is present but it is electrically inert, that is "somehow" hypothetically of-course, it is electrically open-circuited.

2-3-1 Stator Single-Phase Magnetizing Inductance $L_{m,\text{1-phase}}$

As shown in Fig. 2-3a, hypothetically exciting only phase-a (made possible only if the neutral is accessible) by a current i_a results in two equivalent flux components represented in Fig. 2-3b: (1) *magnetizing* flux which crosses the air gap and links with other stator phases and the rotor, and (2) the leakage flux which links phase-a only. Therefore, the self-inductance of a stator phase winding is

$$L_{s,\text{self}} = \frac{\lambda_a}{i_a}\bigg|_{i_a\,\text{only}} = \underbrace{\frac{\lambda_{a,\text{leakage}}}{i_a}}_{L_{\ell s}} + \underbrace{\frac{\lambda_a,\,\text{magnetizing}}{i_a}}_{L_{m,\text{1-phase}}}. \qquad (2\text{-}6a)$$

Therefore,

$$L_{s,\text{self}} = L_{\ell s} + L_{m,\text{1-phase}}. \qquad (2\text{-}6b)$$

It requires no-load and blocked-rotor tests to estimate the leakage inductance $L_{\ell s}$, but the single-phase magnetizing inductance $L_{m,\text{1-phase}}$

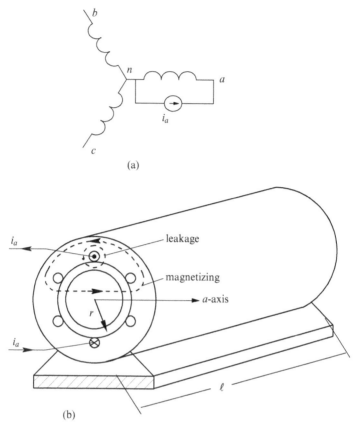

Fig. 2-3 Single-phase magnetizing inductance $L_{m,1\text{-}phase}$ and leakage inductance L_{ls}.

can be easily calculated by equating the energy storage in the air gap to $\frac{1}{2}Li^2$:

$$L_{m,1\text{-phase}} = \frac{\pi \mu_o\, r\ell}{\ell_g}\left(\frac{N_s}{p}\right)^2, \tag{2-7}$$

where r is the mean radius at the air gap, ℓ is the length of the rotor along its shaft axis, N_s equals the number of turns per phase, and p equals the number of poles.

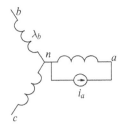

Fig. 2-4 Mutual inductance L_{mutual}.

2-3-2   Stator Mutual-Inductance L_{mutual}

As shown in Fig. 2-4, the mutual-inductance L_{mutual} between two stator phases can be calculated by hypothetically exciting phase-a by i_a and calculating the flux linkage with phase-b

$$L_{\text{mutual}} = \left.\frac{\lambda_b}{i_a}\right|_{i_b, i_c = 0, \text{rotor open}} . \tag{2-8}$$

Note that only the magnetizing flux (not the leakage flux) produced by i_a links the phase-b winding. The current i_a produces a sinusoidal flux-density distribution in the air gap, and the two windings are sinusoidally distributed. Therefore, the flux linking phase-b winding due to i_a can be shown to be the magnetic flux linkage of phase-a winding times the cosine of the angle between the two windings (which in this case is 120°):

$$\lambda_{b, \text{due to } i_a} = \cos(120^o) \lambda_{a, \text{magnetizing due to } i_a} \tag{2-9a}$$

$$= -\frac{1}{2} \lambda_{a, \text{magnetizing due to } i_a}. \tag{2-9b}$$

Therefore, in Eq. (2-8), using Eq. (2-6a) and Eq. (2-9b),

$$L_{\text{mutual}} = -\frac{1}{2} L_{m, \text{1-phase}}. \tag{2-10}$$

The same mutual inductance exists between phase-a and phase-c, and between phase-b and phase-c.

The expression for the mutual inductance can also be derived from energy storage considerations (see Problem 2-2).

2-3-3   Per-Phase Magnetizing-Inductance L_m

Under the condition that the rotor is open-circuited, and all three phases are excited in Fig. 2-2b such that the sum of the three phase currents is zero as given by Eq. (2-5),

$$\lambda_{a,\text{magnetizing}}\Big|_{\substack{\text{(rotor open-circuited)} \\ i_a+i_b+i_c=0}} = L_{m,\text{1-phase}}\, i_a + L_{\text{mutual}}\, i_b + L_{\text{mutual}}\, i_c. \tag{2-11}$$

Using Eq. (2-10) for L_{mutual}, and from Eq. (2-5) replacing $(-i_b - i_c)$ by i_a in Eq. (2-11),

$$L_m = \frac{\lambda_{a,\text{magnetizing}}}{i_a}\Bigg|_{i_a+i_b+i_c=0,\text{rotor open}} = \frac{3}{2} L_{m,\text{1-phase}}. \tag{2-12}$$

Using Eq. (2-7),

$$L_m = \frac{3}{2}\frac{\pi\mu_o r\ell}{\ell_g}\left(\frac{N_s}{p}\right)^2. \tag{2-13}$$

Note that the single-phase magnetizing inductance $L_{m,\text{1-phase}}$ does not include the effect of mutual coupling from the other two phases, whereas the per-phase magnetizing-inductance L_m in Eq. (2-13) does. Hence, L_m is 3/2 times $L_{m,\text{1-phase}}$.

2-3-4   Stator-Inductance L_s

Due to all three stator currents (not including the flux linkage due to the rotor currents), the total flux linkage of phase-a can be expressed as

$$\lambda_a\big|_{\text{rotor-open}} = \lambda_{a,\text{leakage}} + \lambda_{a,\text{magnetizing}}$$
$$= L_{\ell s}i_a + L_m i_a \tag{2-14}$$
$$= L_s i_a,$$

where the stator-inductance L_s is

$$L_s = L_{\ell s} + L_m. \tag{2-15}$$

2-4 EQUIVALENT WINDINGS IN A SQUIRREL-CAGE ROTOR

For developing equations for dynamic analysis, we will replace the squirrel cage on the rotor by a set of three sinusoidally distributed phase windings. The number of turns in each phase of these equivalent rotor windings can be selected arbitrarily. However, the simplest, hence an obvious choice, is to assume that each rotor phase has N_s turns (similar to the stator windings), as shown in Fig. 2-5a. The voltages and currents in these windings are defined in Fig. 2-5b, where the dotted connection to the rotor-neutral is redundant for the following reason: In a balanced rotor, all the bar currents sum to zero at any instant of time (equal currents in either direction). Therefore, in Fig. 2-5b, the three rotor phase currents add up to zero at any instant of time

$$i_A(t)+i_B(t)+i_C(t)=0. \qquad (2\text{-}16)$$

Note that similar to the stator windings, a positive current into a rotor winding produces flux lines in the radially outward direction along its magnetic axis.

2-4-1 Rotor-Winding Inductances (Stator Open-Circuited)

The magnetizing flux produced by each rotor equivalent winding has the same magnetic path in crossing the air gap and the same number of turns as the stator phase windings. Hence, each rotor phase has the same magnetizing inductance $L_{m,\text{1-phase}}$ as the magnetic flux produced by the stator phase winding, although its leakage inductance $L_{\ell r}$ may be different than $L_{\ell s}$. Similarly, L_{mutual} between the two rotor phases would be the same as that between two stator phases. The above equalities also imply that the per-phase magnetizing-inductance L_m in the rotor circuit (under the condition that at any time, $i_A + i_B + i_C = 0$) is the same as that in the stator

$$L_m = \frac{3}{2}L_{m,\text{1-phase}} \qquad (2\text{-}17)$$

and

$$L_r = L_{\ell r} + L_m. \qquad (2\text{-}18)$$

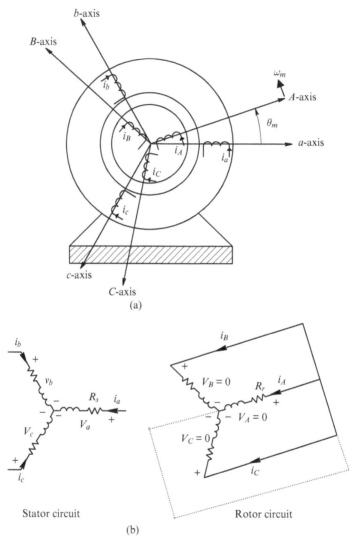

Fig. 2-5   Rotor circuit represented by three-phase windings.

Note that with the choice of the same number of turns in the equiva-lent three-phase rotor windings as in the stator windings, the rotor leakage inductance $L_{\ell r}$ in Eq. (2-18) is the same as $L'_{\ell r}$ in the per-phase, steady-state equivalent circuit of an induction motor. The same applies to the resistances of these equivalent rotor windings, that is, $R_r = R'_r$.

2-5 MUTUAL INDUCTANCES BETWEEN THE STATOR AND THE ROTOR PHASE WINDINGS

If $\theta_m = 0$ in Fig. 2-5a so that the magnetic axes of stator phase-a is aligned with the rotor phase-A, the mutual inductance between the two is at its positive peak and equals $L_{m,\text{1-phase}}$. At any other position of the rotor (including $\theta_m = 0$), this mutual inductance between the stator phase-a and the rotor phase-A can be expressed as

$$L_{aA} = L_{m,\text{1-phase}} \cdot \cos\theta_m. \qquad (2\text{-}19)$$

Similar expressions can be written for mutual inductances between any of the stator phases and any of the rotor phases (see Problem 2-3). Eq. (2-19) shows that the mutual inductance and hence the flux linkages between the stator and the rotor phases vary with position θ_m as the rotor turns.

2-6 REVIEW OF SPACE VECTORS

At any instant of time, each phase winding produces a sinusoidal flux-density distribution (or mmf) in the air gap, which can be represented by a space vector (of the appropriate length) along the magnetic axis of that phase (or opposite to, if the phase current at that instant is negative). These mmf space vectors are $\vec{F}_a^a(t)$, $\vec{F}_b^a(t)$, and $\vec{F}_c^a(t)$, as shown in Fig. 2-6a, with an arrow ("$\rightarrow$") on top of an instantaneous quantity

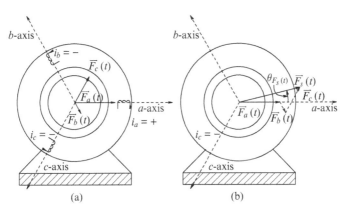

Fig. 2-6 Space vector representation of mmf.

where the superscript "*a*" indicates that the space vectors are expressed as complex numbers, with the stator *a*-axis chosen as the reference axis with an angle of 0°. Assuming that there is no magnetic saturation, the resultant mmf distribution in the air gap due to all three phases at that instant can simply be represented, using vector addition, by the resultant space vector shown in Fig. 2-6b, where the subscript "*s*" represents the combined stator quantities:

$$\vec{F}_s^a(t) = \vec{F}_a^a(t) + \vec{F}_b^a(t) + \vec{F}_c^a(t). \tag{2-20}$$

The earlier explanation provides a physical basis for understanding space vectors. We should note that unlike phasors, space vectors are also applicable under dynamic conditions.

It is easy to visualize the use of space vectors to represent field distributions (F, B, H), which are distributed sinusoidally in the air gap at any instant of time. However, unlike the field quantities, the currents, the voltages, and the flux linkages of phase windings are treated as terminal quantities. The resultant current, voltage, and flux linkage space vectors for the stator are calculated by multiplying instantaneous phase values by the stator winding orientations shown in Fig. 2-7a:

$$\vec{i}_s^a(t) = i_a(t)e^{j0} + i_b(t)e^{j2\pi/3} + i_c(t)e^{j4\pi/3} = \hat{I}_s(t)e^{j\theta_{is}(t)} \tag{2-21}$$

$$\vec{v}_s^a(t) = v_a(t)e^{j0} + v_b(t)e^{j2\pi/3} + v_c(t)e^{j4\pi/3} = \hat{V}_s(t)e^{j\theta_{vs}(t)} \tag{2-22}$$

and

$$\vec{\lambda}_s^a(t) = \lambda_a(t)e^{j0} + \lambda_b(t)e^{j2\pi/3} + \lambda_c(t)e^{j4\pi/3} = \hat{\lambda}_s(t)e^{j\theta_{\lambda s}(t)}. \tag{2-23}$$

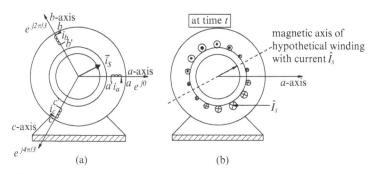

(a)   (b)

Fig. 2-7 Physical interpretation of stator current space vector.

The stator current space vector $\vec{i}_s^{\,a}(t)$ lends itself to a following very useful and simple physical interpretation shown by Fig. 2-7b, noting that in Eq. (2-20), $\vec{F}_s^a(t) = (N_s/p)\vec{i}_s^{\,a}(t)$:

At a time instant t, the three stator phase currents in Fig. 2-7a result in the same mmf acting on the air gap (hence the same flux-density distribution) as that produced by $\vec{i}_s^{\,a}\left(= \hat{I}_s e^{j\theta_{is}}\right)$, that is, by a current equal to its peak value $\hat{I}_s$ flowing through a hypothetical sinusoidally distributed winding shown in Fig. 2-7b, with its magnetic axis oriented at $\theta_{i_s}(=\theta_{F_s})$. This hypothetical winding has the same number of turns N_s sinusoidally-distributed as any of the phase windings.

The earlier physical explanation not only permits the stator current space vector to be visualized, but it also simplifies the derivation of the electromagnetic torque, which can now be calculated on just this single hypothetic winding, rather than having to calculate torques separately on each of the phase windings and then summing them. Similar space vector equations can be written in the rotor circuit with the rotor axis-A as the reference.

2-6-1 Relationship between Phasors and Space Vectors in Sinusoidal Steady State

Under a balanced sinusoidal steady-state condition, the voltage and current phasors in phase-a have the same orientation as the stator voltage and current space vectors at time $t = 0$, as shown for the current in Fig. 2-8; the amplitudes are related by a factor of 3/2:

$$\left.\vec{i}_s^{\,a}\right|_{t=0} = \frac{3}{2}\overline{I}_a \quad \left(\hat{I}_s = \frac{3}{2}\hat{I}_a\right). \qquad (2\text{-}24)$$

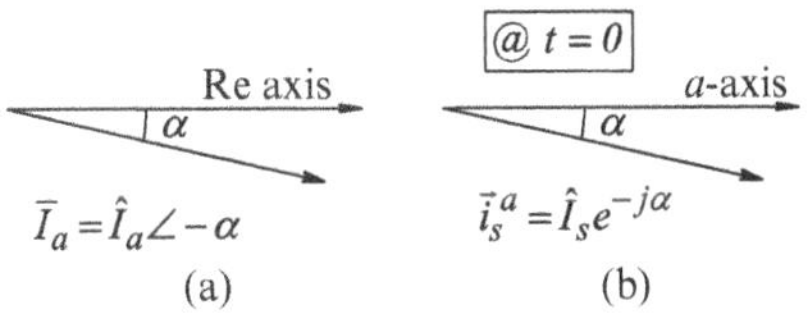

Re axis α $\overline{I}_a = \hat{I}_a \angle -\alpha$ (a)

@ $t = 0$ a-axis α $\vec{i}_s^{\,a} = \hat{I}_s e^{-j\alpha}$ (b)

Fig. 2-8 Relationship between space vector and phasor in sinusoidal steady state.

This relationship is very useful because in our dynamic analyses, we often begin with the induction machine initially operating in a balanced, sinusoidal steady state. (See Problem 2-5.)

2-7   FLUX LINKAGES

In this section, we will develop equations for stator and rotor flux linkages in terms of currents. We will begin by assuming the stator and the rotor to be open-circuited, one at a time. Then, by superposition, based on the assumption of magnetic material in its linear range, we will be able to obtain flux linkages when the stator and the rotor currents are simultaneously present.

2-7-1   Stator Flux Linkage (Rotor Open-Circuited)

In accordance with the Kirchhoff's current law, the currents in the stator windings sum to zero. Initially, we will assume that the rotor is "somehow" open-circuited. Using Eq. (2-14) and Eq. (2-15), writing the flux-linkage equation for each phase and multiplying each equation with its winding orientation

$$[\lambda_{a,i_s}(t) = L_{\ell s}i_a(t) + L_m i_a(t)] \times e^{j0} \tag{2-25a}$$

$$[\lambda_{b,i_s}(t) = L_{\ell s}i_b(t) + L_m i_b(t)] \times e^{j2\pi/3} \tag{2-25b}$$

and

$$[\lambda_{c,i_s}(t) = L_{\ell s}i_c(t) + L_m i_c(t)] \times e^{j4\pi/3}. \tag{2-25c}$$

Using Eq. (2-25a through c) into Eq. (2-23) (where the stator flux linkage due to the rotor currents is not included), the stator flux linkage space vector is

$$\vec{\lambda}^a_{s,i_s}(t) = \underbrace{L_{\ell s}\,\vec{i}^a_s(t)}_{\text{due to leakage flux}} + \underbrace{L_m\,\vec{i}^a_s(t)}_{\text{due to magnetizing flux}} = L_s\,\vec{i}^a_s(t) \quad \text{(rotor open).} \tag{2-26}$$

As in the case of stator current and voltage space vectors, the projection of the stator flux-linkage space vector along a phase axis, multiplied by a factor of 2/3, equals the flux linkage of that phase.

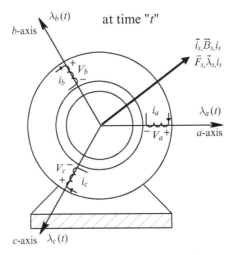

Fig. 2-9 All stator space vectors are collinear (rotor open-circuited).

We have seen earlier that $\vec{i}_s^a(t)$ and $\vec{F}_s^a(t)$ space vectors are collinear, as shown in Fig. 2-9; they are related by a constant. Collinear with $\vec{F}_s^a(t)$, related by a constant μ_0/ℓ_g, is the $\vec{B}_{s,i_s}^a(t)$ space vector, which represents the flux density distribution due the stator currents only, "cutting" the stator conductors. Similarly, the stator flux linkage $\vec{\lambda}_{s,i_s}^a(t)$ in Fig. 2-9 (not including the flux linkage due to the rotor currents) is related to $\vec{i}_s^a(t)$ by a constant L_s as shown by Eq. (2-26). Therefore, all the field quantities, with the rotor open-circuited are collinear, as shown in Fig. 2-9. Note that the superscript "a" is not used while drawing the various space vectors; it needs to be used only while expressing them mathematically, as defined with respect a reference axis, which here is phase-a magnetic axis.

2-7-2 Rotor Flux Linkage (Stator Open-Circuited)

The currents in the rotor equivalent windings sum to zero, as expressed by Eq. (2-16). Assuming that the rotor "somehow" has currents while the stator is open-circuited, by analogy, we can write the expression for the rotor flux linkage space vector as

$$\vec{\lambda}_{r,i_r}^A(t) = \underbrace{L_{\ell r}\,\vec{i}_r^A(t)}_{\text{due to leakage flux}} + \underbrace{L_m\vec{i}_r^A(t)}_{\text{due to magnetizing flux}} = L_r\,\vec{i}_r^A(t) \quad \text{(stator open)}, \quad (2\text{-}27)$$

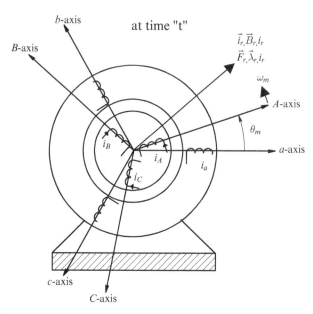

Fig. 2-10   All rotor space vectors are collinear (stator open-circuited).

where the superscript "A" indicates that the rotor phase-A axis is chosen as the reference axis with an angle of $0°$, and $L_r = L_{\ell r} + L_m$. Similar to the stator case, all the field quantities with the stator open circuited are collinear, as shown in Fig. 2-10.

2-7-3   Stator and Rotor Flux Linkages (Simultaneous Stator and Rotor Currents)

When the stator and the rotor currents are present simultaneously, the flux linking any of the stator phases is due to the stator currents as well as the mutual magnetizing flux due to the rotor currents. The magnetizing flux density space vectors in the air gap due to the stator and the rotor currents add up as vectors when these currents are simultaneously present. Therefore, the stator flux linkage, including the leakage flux due to the stator currents can be obtained using Eq. (2-26) and Eq. (2-27) as

$$\vec{\lambda}_s^a(t) = L_s \vec{i}_s^a(t) + L_m \vec{i}_r^a(t), \tag{2-28}$$

where the rotor current space vector is also defined with respect to the stator phase-*a* axis.

Similarly in the rotor circuit, we can write

$$\vec{\lambda}_r^A(t) = L_m \vec{i}_s^A(t) + L_r \vec{i}_r^A(t), \tag{2-29}$$

where the stator current space vector is also defined with respect to the rotor phase-*A* axis.

2-8 STATOR AND ROTOR VOLTAGE EQUATIONS IN TERMS OF SPACE VECTORS

The individual phase equations can be combined to obtain the space vector equation as follows:

$$\left[v_a(t) = R_s i_a(t) + \frac{d}{dt}\lambda_a(t) \right] \times e^{j0} \tag{2-30a}$$

$$\left[v_b(t) = R_s i_b(t) + \frac{d}{dt}\lambda_b(t) \right] \times e^{j2\pi/3} \tag{2-30b}$$

and

$$\left[v_c(t) = R_s i_c(t) + \frac{d}{dt}\lambda_c(t) \right] \times e^{j4\pi/3}. \tag{2-30c}$$

Adding the above three equations and applying the definitions of space vectors, the stator equation can be written as

$$\vec{v}_s^a(t) = R_s \vec{i}_s^a(t) + \frac{d}{dt}\vec{\lambda}_s^a(t). \tag{2-31}$$

Similar to the development in the stator circuit, in the rotor circuit

$$\underbrace{\vec{v}_r^A(t)}_{=0} = R_r \vec{i}_r^A(t) + \frac{d}{dt}\vec{\lambda}_r^A(t), \tag{2-32}$$

where in a squirrel-cage rotor, all the equivalent phase voltages are individually zero and $\vec{v}_r^A(t) = 0$.

2-9 MAKING THE CASE FOR A *dq*-WINDING ANALYSIS

At this point, we should assess how far we have come. The use of space vectors has very quickly allowed us to express the stator and the rotor flux linkages (Eq. 2-28 and Eq. 2-29), which in a compact form include mutual coupling between the six windings: three on the stator and three on the equivalent rotor. In terms of phase quantities of an induction machine, we have developed voltage equations for the rotor and the stator, expressed in a compact space vector form (Eq. 2-31 and Eq. 2-32). These voltage equations include the time derivatives of flux linkages that depend on the rotor position. This dependence can be seen if we examine the flux linkage equations by expressing them with current space vectors defined with respect to their own reference axes in Fig. 2-5 as

$$\vec{i}_r^a(t) = \vec{i}_r^A(t)e^{j\theta_m} \tag{2-33}$$

and

$$\vec{i}_s^A(t) = \vec{i}_s^a(t)e^{-j\theta_m}. \tag{2-34}$$

Using the above two equations in the flux linkage equations (Eq. 2-28 and Eq. 2-29),

$$\vec{\lambda}_s^a(t) = L_s\vec{i}_s^a(t) + L_m\vec{i}_r^A(t)e^{j\theta_m} \tag{2-35}$$

and

$$\vec{\lambda}_r^A(t) = L_m\vec{i}_s^a(t)e^{-j\theta_m} + L_r\vec{i}_r^A(t). \tag{2-36}$$

The flux linkage equations in the above form clearly show their dependence on the rotor position θ_m for given values of the stator and the rotor currents at any instant of time. For this reason, the voltage equations in phase quantities, expressed in a space vector form by Eq. (2-31) and Eq. (2-32), which include the time derivatives of flux linkages, are complicated to solve. It is possible to make these equations simpler by using a transformation called *dq* transformation, which is the topic of the next chapter.

The earlier argument alone on the basis of simplifying the equations, especially in the age of fast (and faster!) computers is not sufficient to search for an alternative, such as the *dq* winding analysis. The power of the *dq* winding analysis lies in the fact that it allows the torque and the flux in the machine to be controlled independently under dynamic conditions, which is not clear in our foregoing analysis based on the phase (*a-b-c*) quantities. An obvious question at this point is if the analysis in this chapter has been a waste. The answer is a resounding "no." We will use every bit of the analysis in this chapter to carry out the *dq* analysis in the next chapter.

Before we embark on the *dq* analysis in the next chapter, we will further look at the analysis of an induction machine in phase quantities by means of the following examples.

EXAMPLE 2-1

First take a two coupled-coil system, one on the stator and the other on the rotor. Derive the electromagnetic torque expression by energy considerations and then generalize it in terms of three-phase stator and rotor currents.

Solution

Neglecting losses, the differential electrical input energy, mechanical energy output, and the stored field energy can be written as follows:

$$dW_{in} = dW_{mech} + dW_{mag}. \qquad (2\text{-}37)$$

For coupled two-coil system of coils 1 and 2, the differential electrical input energy is as follows:

$$
\begin{aligned}
dW_{in} &= v_1 i_1 dt + v_2 i_2 dt \\
&= i_1 d\lambda_1 + i_2 d\lambda_2 \\
&= i_1 d(L_{11} i_1 + L_{12} i_2) + i_2 d(L_{12} i_1 + L_{22} i_2) \qquad (2\text{-}38) \\
&= L_{11} i_1 di_1 + L_{12} i_1 di_2 + i_1^2 dL_{11} + i_1 i_2 dL_{12} + L_{22} i_2 di_2 \\
&\quad + L_{12} i_2 di_1 + i_2^2 dL_{22} + i_1 i_2 dL_{12}.
\end{aligned}
$$

(Continued)

The stored magnetic energy is

$$W_{mag} = \frac{1}{2}L_{11}i_1^2 + \frac{1}{2}L_{22}i_2^2 + L_{12}i_1i_2. \qquad (2\text{-}39)$$

Therefore, the differential increase in the stored magnetic energy is

$$dW_{mag} = L_{11}i_1di_1 + L_{12}i_1di_2 + \frac{1}{2}i_1^2dL_{11} + L_{22}i_2di_2$$
$$+ L_{12}i_2di_1 + \frac{1}{2}i_2^2dL_{22} + i_1i_2dL_{12} \qquad (2\text{-}40)$$

From Eq. (2-37), Eq. (2-38), and Eq. (2-40),

$$T_{em} = \frac{1}{2}i_1^2\frac{dL_{11}}{d\theta_m} + \frac{1}{2}i_2^2\frac{dL_{22}}{d\theta_m} + i_1i_2\frac{dL_{12}}{d\theta_m}. \qquad (2\text{-}41)$$

In terms of a matrix equation, Eq. (2-41) can be written as

$$T_{em} = \frac{1}{2}[i_1 \quad i_2]\frac{d}{\theta_m}\begin{bmatrix} L_{11} & L_{12} \\ L_{12} & L_{22} \end{bmatrix}\begin{bmatrix} i_1 \\ i_2 \end{bmatrix}.$$
$$= \frac{1}{2}[i]^t\frac{d}{\theta_m}[L][i] \qquad (2\text{-}42)$$

Equation (2-42) can be generalized to six windings, *a-b-c* on the stator, and *A-B-C* on the rotor, for a p-pole machine as follows:

$$T_{em} = \frac{p}{2}\frac{1}{2}[i]^t\frac{d}{\theta_m}[L][i], \qquad (2\text{-}43)$$

where $\theta_m = (p/2)\theta_{mech}$.

EXAMPLE 2-2

For the "test" machine given in Chapter 1, simulate the induction machine start-up in MATLAB using the equations derived in this chapter from a completely powered-down state with no external load connected to it. Plot rotor speed, electromagnetic torque, and stator and rotor currents.

Solution

See the complete results, including the computer files, in Appendix 2-A in the accompanying website.

EXAMPLE 2-3

Verify the results in Example 2-2 by simulations in Simulink.

Solution

See the complete results, including the simulation files, in Appendix 2-B in the accompanying website.

2-10 SUMMARY

In this chapter, we have briefly reviewed the sinusoidally distributed windings and then calculated their inductances for developing equations for induction machines in phase (a-b-c) quantities. The development of these equations is assisted by space vectors, which are briefly reviewed. The analysis in this chapter establishes the framework and the rationale for the dq windings-based analysis of induction machines under dynamic conditions carried out in the next chapter.

REFERENCE

1. N. Mohan, *Electric Machines and Drives: A First Course*, Wiley, Hoboken, NJ, 2011. http://www.wiley.com/college/mohan.

PROBLEMS

2-1 Derive Eq. (2-7) for $L_{m,1\text{-phase}}$.

2-2 Derive the expression for L_{mutual} in Eq. (2-8) by energy storage considerations. Hint: Assume that only the stator phases a and b are excited so that $i_b = -i_a$. To keep this derivation general, begin by assuming an arbitrary angle θ between the magnetic axes of the two windings.

2-3 Write the expressions for L_{kJ} as functions of θ_m, where $k \equiv a,b,c$ and $J \equiv A,B,C$.

2-4 Calculate L_s, L_r, and L_m for the "test" motor described in Chapter 1.

2-5 A motor with the following nameplate data is operating in a balanced sinusoidal steady state under its rated condition (with rated voltages applied to it and it is loaded to its rated torque). Assume that the voltage across phase-a is at its positive peak at $t = 0$. (a) Obtain at time $t = 0$, $\vec{v}_s(0)$, $\vec{i}_s(0)$, and $\vec{i}_r(0)$, and (b) express phase voltages and currents as functions of time.

Nameplate Data

Power:	3 HP/2.4 kW
Voltage:	460 V (L-L, rms)
Frequency:	60 Hz
Phases:	3
Full Load Current:	4 A
Full-Load Speed:	1750 rpm
Full-Load Efficiency:	88.5%
Power Factor:	80.0%
Number of Poles:	4

Per-Phase Motor Circuit Parameters:

$$R_s = 1.77\,\Omega$$
$$R_r = 1.34\,\Omega$$
$$X_{\ell s} = 5.25\,\Omega\ (\text{at }60\text{ Hz})$$

$X_{\ell r} = 4.57\,\Omega\,(\text{at 60 Hz})$
$X_m = 139.0\,\Omega\,(\text{at 60 Hz})$
Full-Load Slip = 1.72%

The iron losses are specified as 78W and the mechanical (friction and windage) losses are specified as 24W. The inertia of the machine is given. Assuming that the reflected load inertia is approximately the same as the motor inertia, the total equivalent inertia of the system is $J_{eq} = 0.025\,\text{kg}\cdot\text{m}^2$.

2-6 At an instant of time in an induction machine, hypothetically assume that the stator currents $i_a = 10\text{A}, i_b = -3\text{A}, i_c = -7\text{A}$, and the rotor currents $i_A = 3\text{A}, i_B = -1\text{A}, i_C = -2\text{A}$. Calculate $\vec{\lambda}_s^a\big|_{\vec{i}_s}(t)$, $\vec{\lambda}_r^A\big|_{\vec{i}_r}(t)$, $\vec{\lambda}_s^a(t)$, and $\vec{\lambda}_r^A(t)$ in terms of machine inductances L_m, L_s, and L_r, if the rotor angle θ_m has the following values: (a) $0°$ and (b) $30°$.

2-7 Write the expression for the stator phase-a flux linkage in terms of three stator and three rotor phase currents and the appropriate inductances, for a rotor position of θ_m. Repeat this for the other stator and rotor phases.

2-8 Show that Eq. (2-28) and Eq. (2-29) can be written with respect to any arbitrary axis, rather than a-axis or A-axis.

2-9 Show the intermediate steps in generalizing Eq. (2-42) to Eq. (2-43).

3 Dynamic Analysis of Induction Machines in Terms of *dq* Windings

3-1 INTRODUCTION

In this chapter, we will develop equations to analyze induction machine operation under dynamic conditions. We will make use of space vectors as intermediary in transforming *a-b-c* phase winding quantities into equivalent *dq*-winding quantities that we will use for dynamic (non-steady state) analysis. We will see in later chapters the benefits of *d*- and *q*-axis analysis in controlling ac machines.

3-2 *dq* WINDING REPRESENTATION

We studied in the previous chapter that the stator and the rotor flux linkages $\vec{\lambda}_s^a(t)$ and $\vec{\lambda}_r^a(t)$ depend on the rotor angle θ_m because the mutual inductances between the stator and the rotor windings are position dependent. The main reason for the *d*- and *q*-axis analysis in machines like the induction machines is to control them properly, for example, using vector control principles. In most textbooks, this analysis is discussed as a mathematical transformation called the Park's transformation. In this chapter, we will take a physical approach to this transformation, which is much easier to visualize and arrive at identical results.

Advanced Electric Drives: Analysis, Control, and Modeling Using MATLAB/Simulink®, First Edition. Ned Mohan.

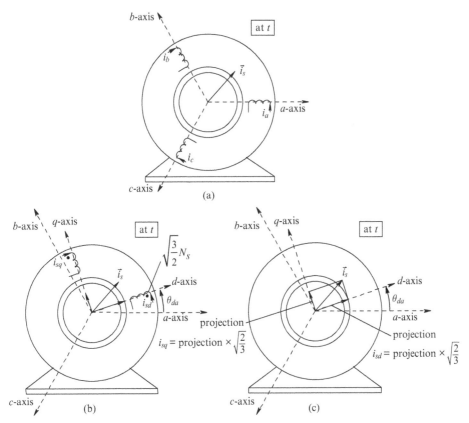

Fig. 3-1 Representation of stator mmf by equivalent *dq* windings.

3-2-1 Stator *dq* Winding Representation

In Fig. 3-1a at time *t*, phase currents $i_a(t)$, $i_b(t)$, and $i_c(t)$ are represented by a stator current space vector $\vec{i}_s(t)$. A collinear magnetomotive force (mmf) space vector $\vec{F}_s(t)$ is related to $\vec{i}_s(t)$ by a factor of (N_s/p), where N_s equals the number of turns per phase and p equals the number of poles:

$$\vec{i}_s^{\,a}(t) = i_a(t) + i_b(t)e^{j2\pi/3} + i_c(t)e^{j4\pi/3} \tag{3-1}$$

and

$$\vec{F}_s^{\,a}(t) = \frac{N_s}{p}\,\vec{i}_s^{\,a}(t). \tag{3-2}$$

We should note that the space vector $\vec{i}_s(t)$ in Fig. 3-1 is written without a superscript "a." The reason is that a reference axis is needed *only* to express it mathematically by means of complex numbers. However, $\vec{i}_s(t)$ in Fig. 3-1 depends on the instantaneous values of phase currents and is independent of the choice of the reference axis to draw it.

In the previous course for analyzing ac machines under balanced sinusoidal steady-state conditions, we replaced the three windings by a single hypothetical equivalent winding that produced the same mmf distribution in the air gap. This single winding was sinusoidally distributed with the same number of turns N_s (as any phase winding), with its magnetic axis along the stator current space vector and a current $\hat{I}_s$ (peak value of $\vec{i}_s$) flowing through it.

However, for dynamic analysis and control of ac machines, we need two orthogonal windings such that the torque and the flux within the machine can be controlled independently. At any instant of time, the air gap mmf distribution by three phase-windings can also be produced by a set of two orthogonal windings shown in Fig. 3-1b, each sinusoidally distributed with $\sqrt{3/2}N_s$ turns: one winding along the d-axis, and the other along the q-axis. The reason for choosing $\sqrt{3/2}N_s$ turns will be explained shortly. This dq winding set may be at any arbitrary angle θ_{da} with respect to the phase-a axis. However, the currents i_{sd} and i_{sq} in these two windings must have specific values, which can be obtained by equating the mmf produced by the dq windings to that produced by the three phase windings and represented by a single winding with N_s turns in Eq. (3-2)

$$\frac{\sqrt{3/2}N_s}{p}(i_{sd}+ji_{sq})=\frac{N_s}{p}\vec{i}_s^{\,d},\qquad(3\text{-}3)$$

where the stator current space vector is expressed using the d-axis as the reference axis, hence the superscript "d." Eq. (3-3) results in

$$(i_{sd}+ji_{sq})=\sqrt{\frac{2}{3}}\vec{i}_s^{\,d},\qquad(3\text{-}4)$$

which shows that the dq winding currents are $\sqrt{2/3}$ times the projections of $\vec{i}_s(t)$ vector along the d- and q-axis, as shown in Fig. 3-1c:

$$i_{sd} = \sqrt{2/3} \times \text{projection of } \vec{i}_s(t) \text{ along the } d\text{-axis} \qquad (3\text{-}5)$$

and

$$i_{sq} = \sqrt{2/3} \times \text{projection of } \vec{i}_s(t) \text{ along the } q\text{-axis.} \qquad (3\text{-}6)$$

The factor $\sqrt{2/3}$, reciprocal of the factor $\sqrt{3/2}$ used in choosing the number of turns for the *dq* windings, ensures that the *dq*-winding currents produce the same mmf distribution as the three-phase winding currents.

In Fig. 3-1b, the *d* and the *q* windings are mutually decoupled magnetically due to their orthogonal orientation. Choosing $\sqrt{3/2}N_s$ turns for each of these windings results in their magnetizing inductance to be L_m (same as the per-phase magnetizing inductance in Chapter 2 for three-phase windings with $i_a + i_b + i_c = 0$) for the following reason: the inductance of a winding is proportional to the square of the number of turns and therefore, the magnetizing inductance of any *dq* winding (noting that there is no mutual inductance between the two orthogonal windings) is

$$\begin{aligned} dq \text{ winding magnetizing inductance} &= (\sqrt{3/2})^2 L_{m,1\text{-phase}} \\ &= (3/2)L_{m,1\text{-phase}} \qquad (3\text{-}7) \\ &= L_m \text{ (using Eq. 2-12).} \end{aligned}$$

Each of these equivalent windings has a resistance R_s and a leakage inductance $L_{\ell s}$, similar to the *a-b-c* phase windings (see Problem 3-1). In fact, if a three-phase machine were to be converted to a two-phase machine using the same stator shell (but the windings could be different) to deliver the same power output and speed, we will choose the number of turns in the each of the two-phase windings to be $\sqrt{3/2}N_s$.

3-2-2 Rotor *dq* Windings (Along the Same *dq*-Axes as in the Stator)

The rotor mmf space vector $\vec{F}_r(t)$ is produced by the combined effect of the rotor bar currents, or by the three equivalent phase windings, each with N_s turns, as shown in Fig. 3-2 (short-circuited in a squirrel-cage rotor). The phase currents in these equivalent rotor phase windings can be represented by a rotor current space vector, where

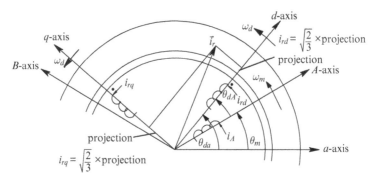

Fig. 3-2   Representation of rotor mmf by equivalent *dq* winding currents.

$$\vec{i}_r^{\,A}(t) = i_A(t) + i_B(t)e^{j2\pi/3} + i_C(t)e^{j4\pi/3}, \qquad (3\text{-}8)$$

where

$$\vec{i}_r^{\,A}(t) = \frac{\vec{F}_r^{\,A}(t)}{N_s / p}. \qquad (3\text{-}9)$$

The mmf $\vec{F}_r(t)$ and the rotor current $\vec{i}_r(t)$ in Fig. 3-2 can also be produced by the components $i_{rd}(t)$ and $i_{rq}(t)$ flowing through their respective windings as shown. (Note that the *d*- and the *q*-axis are the same as those chosen for the stator in Fig. 3-1. Otherwise, all benefits of the *dq*-analysis will be lost.) Similar to the stator case, each of the *dq* windings on the rotor has $\sqrt{3/2}N_s$ turns, and a magnetizing inductance of L_m, which is the same as that for the stator *dq* windings because of the same number of turns (by choice) and the same magnetic path for flux lines. Each of these rotor equivalent windings has a resistance R_r and a leakage inductance $L_{\ell r}$ (equal to R_r' and $L_{\ell r}'$, respectively, in the per-phase equivalent circuit of induction machines in the previous course). The mutual inductance between these two orthogonal windings is zero.

3-2-3   Mutual Inductance between *dq* Windings on the Stator and the Rotor

The equivalent *dq* windings for the stator and the rotor are shown in Fig. 3-3. The mutual inductance between the stator and the rotor *d*-axis windings is equal to L_m due to the magnetizing flux crossing the air gap. Similarly, the mutual inductance between the stator and the rotor *q*-axis

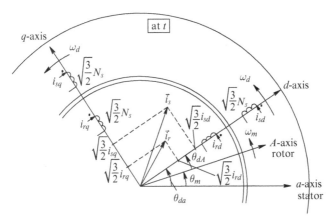

Fig. 3-3 Stator and rotor representation by equivalent *dq* winding currents. The *dq* winding voltages are defined as positive at the dotted terminals. Note that the relative positions of the stator and the rotor current space vectors are not actual, rather only for definition purposes.

windings equals L_m. Out of four *dq* windings, the mutual inductance between any *d*-axis winding with any *q*-axis winding is zero because of their orthogonal orientation, which results in zero mutual magnetic coupling of flux.

3-3   MATHEMATICAL RELATIONSHIPS OF THE *dq* WINDINGS (AT AN ARBITRARY SPEED ω_d)

Next, we will describe relationships between the stator and the rotor quantities and their equivalent *dq* winding components in Fig. 3-3, which in combination produce the same mmf as the actual three phase windings.

It is worth repeating that the space vectors at some arbitrary time *t* in Fig. 3-3 are expressed without a superscript "*a*" or "*A*." The reason is that a reference axis is needed *only* to express them mathematically by means of complex numbers. In other words, these space vectors in Fig. 3-3 would be in the same position, independent of the choice of the reference axis to express them. We should note that the relative position of $\vec{i}_s$ and $\vec{i}_r$ is shown arbitrarily here just for definition purposes (in an induction machine, the angle between $\vec{i}_s$ and $\vec{i}_r$ is very large— more than $145°$).

Hereafter, we will drop the superscript to any space vector expressed using d-*axis as the reference.*

From Fig. 3-3, we note that at time t, the d-axis is shown at an angle θ_{da} with respect to the stator a-axis. Therefore,

$$\vec{i}_s(t) = \vec{i}_s^a(t)e^{-j\theta_{da}(t)}. \tag{3-10}$$

Substituting for $\vec{i}_s^a$ from Eq. (3-1),

$$\vec{i}_s(t) = i_a e^{-j\theta_{da}} + i_b e^{-j(\theta_{da}-2\pi/3)} + i_c e^{-j(\theta_{da}-4\pi/3)}. \tag{3-11}$$

Equating the real and imaginary components on the right side of Eq. (3-11) to i_{sd} and i_{sq} in Eq. (3-4)

$$\begin{bmatrix} i_{sd}(t) \\ i_{sq}(t) \end{bmatrix} = \sqrt{\frac{2}{3}} \underbrace{\begin{bmatrix} \cos(\theta_{da}) & \cos\left(\theta_{da} - \dfrac{2\pi}{3}\right) & \cos\left(\theta_{da} - \dfrac{4\pi}{3}\right) \\ -\sin(\theta_{da}) & -\sin\left(\theta_{da} - \dfrac{2\pi}{3}\right) & -\sin\left(\theta_{da} - \dfrac{4\pi}{3}\right) \end{bmatrix}}_{[T_s]_{abc \to dq}} \begin{bmatrix} i_a(t) \\ i_b(t) \\ i_c(t) \end{bmatrix},$$

$$\tag{3-12}$$

where $[T_s]_{abc \to dq}$ is the transformation matrix to transform stator a-b-c phase winding currents to the corresponding dq winding currents. This transformation procedure is illustrated by the block diagram in Fig. 3-4a. The same transformation matrix relates the stator flux linkages and the stator voltages in phase windings to those in the equivalent stator dq windings.

A similar procedure to that in the stator case is followed for the rotor where in terms of the phase currents, the rotor current space vector is

$$\vec{i}_r^A(t) = i_A(t) + i_B(t)e^{j2\pi/3} + i_C(t)e^{j4\pi/3}. \tag{3-13}$$

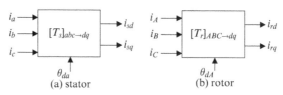

Fig. 3-4   Transformation of phase quantities into *dq* winding quantities.

From Fig. 3-3, we note that at time t, d-axis is at an angle θ_{dA} with respect to the rotor A-axis. Therefore,

$$\vec{i}_r(t) = \vec{i}_r^A(t)e^{-j\theta_{dA}(t)}.\tag{3-14}$$

The currents in the dq rotor windings must be i_{rd} and i_{rq}, where these two current components are $\sqrt{2/3}$ times the projections of $\vec{i}_r(t)$ vector along the d- and q-axis, as shown in Fig. 3-3

$$i_{rd} = \sqrt{2/3} \times \text{projection of } \vec{i}_r(t) \text{ along the } d\text{-axis}\tag{3-15}$$

and

$$i_{rq} = \sqrt{2/3} \times \text{projection of } \vec{i}_r(t) \text{ along the } q\text{-axis.}\tag{3-16}$$

Similar to Eq. (3-12), replacing θ_{da} by θ_{dA}

$$\begin{bmatrix} i_{rd}(t) \\ i_{rq}(t) \end{bmatrix} = \sqrt{\frac{2}{3}} \underbrace{\begin{bmatrix} \cos(\theta_{dA}) & \cos\left(\theta_{dA} - \frac{2\pi}{3}\right) & \cos\left(\theta_{dA} - \frac{4\pi}{3}\right) \\ -\sin(\theta_{dA}) & -\sin\left(\theta_{dA} - \frac{2\pi}{3}\right) & -\sin\left(\theta_{dA} - \frac{4\pi}{3}\right) \end{bmatrix}}_{[T_r]_{ABC \to dq}} \begin{bmatrix} i_A(t) \\ i_B(t) \\ i_C(t) \end{bmatrix},$$

$$\tag{3-17}$$

where $[T_r]_{ABC \to dq}$ is the transformation matrix for the rotor. This transformation procedure is illustrated by the block diagram in Fig. 3-4b, similar to that in Fig. 3-4a. The same transformation matrix relates the rotor flux linkages and the rotor voltages in the equivalent A-B-C windings to those in the equivalent rotor dq windings. Same relationships apply to voltages and flux linkages.

3-3-1 Relating *dq* Winding Variables to Phase Winding Variables

In case of an isolated neutral, where all three phase currents add up to zero at any time, the variables in a-b-c phase windings can be calculated in terms of the dq-winding variables. In Eq. (3-12), we can add a row at the bottom to represent the condition that all three phase currents sum to zero. Inverting the resulting matrix and discarding the last column whose contribution is zero, we obtain the desired relationship

$$\begin{bmatrix} i_a(t) \\ i_b(t) \\ i_c(t) \end{bmatrix} = \sqrt{\frac{2}{3}} \underbrace{\begin{bmatrix} \cos(\theta_{da}) & -\sin(\theta_{da}) \\ \cos\left(\theta_{da} + \frac{4\pi}{3}\right) & -\sin\left(\theta_{da} + \frac{4\pi}{3}\right) \\ \cos\left(\theta_{da} + \frac{2\pi}{3}\right) & -\sin\left(\theta_{da} + \frac{2\pi}{3}\right) \end{bmatrix}}_{[T_s]_{dq \to abc}} \begin{bmatrix} i_{sd} \\ i_{sq} \end{bmatrix}, \tag{3-18}$$

where $[T_s]_{dq \to abc}$ is the transformation matrix in the reverse direction (*dq* to *abc*). A similar transformation matrix $[T_r]_{dq \to ABC}$ for the rotor can be written by replacing θ_{da} in Eq. (3-18) by θ_{dA}.

3-3-2   Flux Linkages of *dq* Windings in Terms of Their Currents

We have a set of four *dq* windings as shown in Fig. 3-3. There is no mutual coupling between the windings on the *d*-axis and those on the *q*-axis. The flux linking any winding is due to its own current and that due to the other winding on the same axis. Let us select the stator *d*-winding as an example. Due to i_{sd}, both the magnetizing flux as well as the leakage flux link this winding. However, due to i_{rd}, only the magnetizing flux (leakage flux does not cross the air gap) links this stator winding. Using this logic, we can write the following flux expressions for all four windings:

Stator Windings

$$\lambda_{sd} = L_s i_{sd} + L_m i_{rd} \tag{3-19}$$

and

$$\lambda_{sq} = L_s i_{sq} + L_m i_{rq}, \tag{3-20}$$

where in Eq. (3-19) and Eq. (3-20), $L_s = L_{\ell s} + L_m$.

Rotor Windings

$$\lambda_{rd} = L_r i_{rd} + L_m i_{sd} \tag{3-21}$$

and

$$\lambda_{rq} = L_r i_{rq} + L_m i_{sq}, \tag{3-22}$$

where in Eq. (3-21) and Eq. (3-22), $L_r = L_{\ell r} + L_m$.

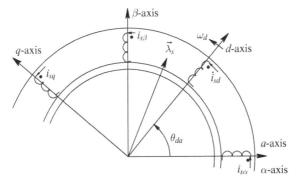

Fig. 3-5   Stator $\alpha\beta$ and *dq* equivalent windings.

3-3-3   *dq* Winding Voltage Equations

Stator Windings

To derive the *dq* winding voltages, we will first consider a set of orthogonal $\alpha\beta$ windings affixed to the stator, as shown in Fig. 3-5, where the α-axis is aligned with the stator *a*-axis. In all windings, the voltage polarity is defined to be positive at the dotted terminal. In $\alpha\beta$ windings, in terms of their variables,

$$v_{s\alpha} = R_s i_{s\alpha} + \frac{d}{dt}\lambda_{s\alpha} \tag{3-23}$$

and

$$v_{s\beta} = R_s i_{s\beta} + \frac{d}{dt}\lambda_{s\beta}. \tag{3-24}$$

The above two equations can be combined by multiplying both sides of Eq. (3-24) by the operator (j) and then adding to Eq. (3-23). In terms of resulting space vectors

$$\vec{v}_{s\_\alpha\beta}^{\,\alpha} = R_s \vec{i}_{s\_\alpha\beta}^{\,\alpha} + \frac{d}{dt}\vec{\lambda}_{s\_\alpha\beta}^{\,\alpha}, \tag{3-25}$$

where $\vec{v}_{s\_\alpha\beta}^{\,\alpha} = v_{s\alpha} + jv_{s\beta}$ and so on. As can be seen from Fig. 3-5, the current, voltage, and flux linkage space vectors with respect to the α-axis are related to those with respect to the *d*-axis as follows:

$$\vec{v}_{s\_\alpha\beta}^{\,\alpha} = \vec{v}_{s\_dq} \cdot e^{j\theta_{da}} \tag{3-26a}$$

$$\vec{i}_{s\_\alpha\beta}^{\,\alpha} = \vec{i}_{s\_dq} \cdot e^{j\theta_{da}} \tag{3-26b}$$

and

$$\vec{\lambda}_{s\_\alpha\beta}^{\alpha} = \vec{\lambda}_{s\_dq} \cdot e^{j\theta_{da}}, \tag{3-26c}$$

where $\vec{v}_{s\_dq} = v_{sd} + jv_{sq}$ and so on. Substituting expressions from Eq. (3-26a through c) into Eq. (3-25),

$$\vec{v}_{s\_dq} \cdot e^{j\theta_{da}} = R_s \vec{i}_{s\_dq} \cdot e^{j\theta_{da}} + \frac{d}{dt}(\vec{\lambda}_{s\_dq} \cdot e^{j\theta_{da}})$$

or

$$\vec{v}_{s\_dq} \cdot \cancel{e^{j\theta_{da}}} = R_s \vec{i}_{s\_dq} \cdot \cancel{e^{j\theta_{da}}} + \frac{d\vec{\lambda}_{s\_dq}}{dt} \cdot \cancel{e^{j\theta_{da}}} + j\underbrace{\frac{d\theta_{da}}{dt}}_{\omega_d} \cdot \vec{\lambda}_{s\_dq} \cdot \cancel{e^{j\theta_{da}}}.$$

Hence,

$$\vec{v}_{s\_dq} = R_s \vec{i}_{s\_dq} + \frac{d}{dt}\vec{\lambda}_{s\_dq} + j\omega_d \vec{\lambda}_{s\_dq}, \tag{3-27}$$

where $(d/dt)\theta_{da} = \omega_d$ is the instantaneous speed (in electrical radians per second) of the *dq* winding set in the air gap, as shown in Fig. 3-3 and Fig. 3-5. Separating the real and imaginary components in Eq. (3-27), we obtain

$$v_{sd} = R_s i_{sd} + \frac{d}{dt}\lambda_{sd} - \omega_d \lambda_{sq} \tag{3-28}$$

and

$$v_{sq} = R_s i_{sq} + \frac{d}{dt}\lambda_{sq} + \omega_d \lambda_{sd}. \tag{3-29}$$

In Eq. (3-28) and Eq. (3-29), the speed terms are the components that are proportional to ω_d (the speed of the *dq* reference frame relative to the actual physical stator winding speed) and to the flux linkage of the orthogonal winding.

Equation (3-28) and Equation (3-29) can be written as follows in a vector form, where each vector contains a pair of variables—the first entry corresponds to the *d*-winding and the second to the *q*-winding:

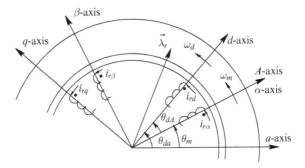

Fig. 3-6 Rotor $\alpha\beta$ and dq equivalent windings.

$$\begin{bmatrix} v_{sd} \\ v_{sq} \end{bmatrix} = R_s \begin{bmatrix} i_{sd} \\ i_{sq} \end{bmatrix} + \frac{d}{dt}\begin{bmatrix} \lambda_{sd} \\ \lambda_{sq} \end{bmatrix} + \omega_d \underbrace{\begin{bmatrix} 0 & -1 \\ 1 & 0 \end{bmatrix}}_{[\mathrm{M_{rotate}}]}\begin{bmatrix} \lambda_{sd} \\ \lambda_{sq} \end{bmatrix}. \tag{3-30}$$

Note that the 2×2 matrix $[\mathrm{M_{rotate}}]$ in Eq. (3-30) in the vector form corresponds to the operator (j) in Eq. (3-27), where $j(= e^{j\pi/2})$ has the role of rotating the space vector $\vec{\lambda}_{s\_dq}$ by an angle of $\pi/2$.

Rotor Windings

An analysis similar to the stator case is carried out for the rotor, where the $\alpha\beta$ windings affixed to the rotor are shown in Fig. 3-6 with the α-axis aligned with rotor A-axis. The d-axis (same as the d-axis for the stator) in this case is at an angle θ_{dA} with respect to the A-axis. Following the procedure for the stator case by replacing θ_{da} by θ_{dA} results in the following equations for the rotor winding voltages

$$v_{rd} = R_r i_{rd} + \frac{d}{dt}\lambda_{rd} - \omega_{dA}\lambda_{rq} \tag{3-31}$$

and

$$v_{rq} = R_r i_{rq} + \frac{d}{dt}\lambda_{rq} + \omega_{dA}\lambda_{rd}, \tag{3-32}$$

where

$$\frac{d}{dt}\theta_{dA} = \omega_{dA}$$

is the instantaneous speed (in electrical radians per second) of the dq winding set in the air gap with respect to the rotor A-axis speed (rotor speed), that is,

$$\omega_{dA} = \omega_d - \omega_m. \tag{3-33}$$

In Eq. (3-33), ω_m is the rotor speed in electrical radians per second. It is related to ω_{mech}, the rotor speed in actual radians per second, by the pole-pairs as follows:

$$\omega_m = (p/2)\omega_{mech}. \tag{3-34}$$

In Eq. (3-31) and Eq. (3-32), the speed terms are the components that are proportional to ω_{dA} (the speed of the dq reference frame relative to the actual physical rotor winding speed) and to the flux linkage of the orthogonal winding.

Equation (3-31) and Equation (3-32) can be written as follows in a vector form, where each vector contains a pair of numbers—the first entry corresponds to the d-axis and the second to the q-axis:

$$\begin{bmatrix} v_{rd} \\ v_{rq} \end{bmatrix} = R_r \begin{bmatrix} i_{rd} \\ i_{rq} \end{bmatrix} + \frac{d}{dt}\begin{bmatrix} \lambda_{rd} \\ \lambda_{rq} \end{bmatrix} + \omega_{dA} \underbrace{\begin{bmatrix} 0 & -1 \\ 1 & 0 \end{bmatrix}}_{[\text{M}_{\text{rotate}}]}\begin{bmatrix} \lambda_{rd} \\ \lambda_{rq} \end{bmatrix}. \tag{3-35}$$

3-3-4 Obtaining Fluxes and Currents with Voltages as Inputs

We can write Eq. (3-30) and Eq. (3-35) in a state space form as follows:

$$\frac{d}{dt}\begin{bmatrix} \lambda_{sd} \\ \lambda_{sq} \end{bmatrix} = \begin{bmatrix} v_{sd} \\ v_{sq} \end{bmatrix} - R_s\begin{bmatrix} i_{sd} \\ i_{sq} \end{bmatrix} - \omega_d \underbrace{\begin{bmatrix} 0 & -1 \\ 1 & 0 \end{bmatrix}}_{[\text{M}_{\text{rotate}}]}\begin{bmatrix} \lambda_{sd} \\ \lambda_{sq} \end{bmatrix} \tag{3-36}$$

and

$$\frac{d}{dt}\begin{bmatrix} \lambda_{rd} \\ \lambda_{rq} \end{bmatrix} = \begin{bmatrix} v_{rd} \\ v_{rq} \end{bmatrix} - R_r\begin{bmatrix} i_{rd} \\ i_{rq} \end{bmatrix} - \omega_{dA} \underbrace{\begin{bmatrix} 0 & -1 \\ 1 & 0 \end{bmatrix}}_{[\text{M}_{\text{rotate}}]}\begin{bmatrix} \lambda_{rd} \\ \lambda_{rq} \end{bmatrix}. \tag{3-37}$$

Assigning $[\lambda_{s\_dq}]$, $[v_{s\_dq}]$, and so on to represent these vectors, Eq. (3-36) and Eq. (3-37) can be written as

$$\frac{d}{dt}[\lambda_{s\_dq}] = [v_{s\_dq}] - R_s[i_{s\_dq}] - \omega_d[\text{M}_{\text{rotate}}][\lambda_{s\_dq}] \tag{3-38}$$

and

$$\frac{d}{dt}[\lambda_{r\_dq}] = [v_{r\_dq}] - R_r[i_{r\_dq}] - \omega_{dA}[\text{M}_{\text{rotate}}][\lambda_{r\_dq}]. \tag{3-39}$$

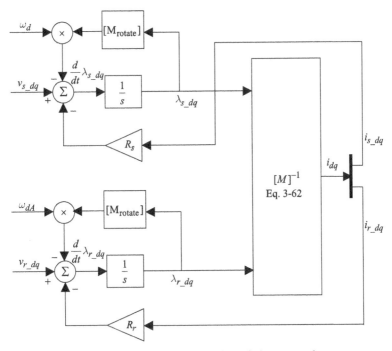

Fig. 3-7 Calculating *dq* winding flux linkages and currents.

Equation (3-38) and Equation (3-39) are represented by a block diagram in Fig. 3-7, where the calculation of *dq* winding currents from flux linkages is formalized in Section 3-9.

3-4   CHOICE OF THE *dq* WINDING SPEED ω_d

It is possible to assume any arbitrary value for the *dq* winding speed ω_d. However, there is one value (out of three) that usually makes sense: $\omega_d = \omega_{syn}$, 0 or ω_m, where ω_{syn} is the synchronous speed in electrical radians per second. The corresponding values for ω_{dA} equal ω_{slip}, $-\omega_m$ or 0, respectively, where $\omega_{slip} = \omega_{syn} - \omega_m$ in electrical radians per second.

Under a balanced sinusoidal steady state, the choice of $\omega_d = \omega_{syn}$ (hence $\omega_{dA} = \omega_{slip}$) results in the hypothetical *dq* windings rotating at the same speed as the field distribution in the air gap. Therefore, all the currents, voltages, and flux linkages associated with the stator and the rotor *dq* windings are dc in a balanced sinusoidal steady state. It is easy to design *PI* controllers for dc quantities, hence ω_{syn} is often the choice for ω_d.

In contrast, choosing $\omega_d = 0$, that is, a stationary *d*-axis (often chosen to be aligned with the *a*-axis of the stator with $\theta_{da} = 0$), leads to the rotor and the stator *dq* winding voltages and currents oscillating at the synchronous frequency in a balanced sinusoidal steady state. The choice of $\omega_d = \omega_m$ results in *dq* winding voltages and currents in the stator and the rotor varying at the slip frequency; this choice is made for analyzing synchronous machines, as we will discuss in Chapter 9.

3-5   ELECTROMAGNETIC TORQUE

3-5-1   Torque on the Rotor *d*-Axis Winding

On the rotor *d*-axis winding, the torque produced is due to the flux density produced by the *q*-axis windings in Fig. 3-8. The peak of the flux density distribution "cutting" the rotor *d* winding due to i_{sq} and i_{rq}, each flowing through $\sqrt{3/2}N_s$ turns of the *q*-axis windings (using Eq. 2-3), is:

$$\hat{B}_{rq} = \frac{\mu_0}{\ell_g}\underbrace{\left(\frac{\sqrt{3/2}N_s}{p}\right)\left(i_{sq} + \frac{L_r}{L_m}i_{rq}\right)}_{\text{mmf}}, \qquad (3\text{-}40)$$

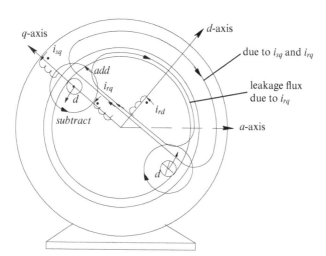

Fig. 3-8   Torque on the rotor *d*-axis.

where the factor L_r/L_m allows us to include both the magnetizing and the leakage flux produced by i_{rq}. Using the torque expression in chapter 10 of Reference [1] used in the previous course, and noting that the current i_{rd} in the rotor d-axis winding flows through $\sqrt{3/2}N_s$ turns, the instantaneous torque on the d-axis rotor winding is (see problem 10-1 in chapter 10 of Reference [1])

$$T_{d,\text{rotor}} = \frac{p}{2}\left(\pi \frac{\sqrt{3/2}N_s}{p} r\ell \hat{B}_{rq}\right) i_{rd}.$$

(3-41)

As shown in Fig. 3-8, this torque on the rotor is counter-clockwise (CCW), hence we will consider it as positive. Substituting for $\hat{B}_{rq}$ from Eq. (3-40) into Eq. (3-41),

$$T_{d,\text{rotor}} = \frac{p}{2}\left(\pi \frac{\mu_0}{\ell_g} r\ell\right)\left(\frac{\sqrt{3/2}N_s}{p}\right)^2\left(i_{sq} + \frac{L_r}{L_m}i_{rq}\right) i_{rd}.$$

(3-42)

Rewriting Eq. (3-42) below, we can recognize L_m from Eq. (2-13)

$$T_{d,\text{rotor}} = \frac{p}{2}\underbrace{\left[\frac{3}{2}\pi \frac{\mu_0}{\ell_g} r\ell\left(\frac{N_s}{p}\right)^2\right]}_{L_m}\left(i_{sq} + \frac{L_r}{L_m}i_{rq}\right) i_{rd}.$$

Hence,

$$T_{d,\text{rotor}} = \frac{p}{2}\underbrace{(L_m i_{sq} + L_r i_{rq})}_{\lambda_{rq}} i_{rd} = \frac{p}{2}\lambda_{rq} i_{rd}.$$

(3-43)

3-5-2 Torque on the Rotor q-Axis Winding

On the rotor q-axis winding, the torque produced is due to the flux density produced by the d-axis windings in Fig. 3-9. This torque on the rotor is clockwise (CW), hence we will consider it as negative. The derivation similar to that of the torque expression on the rotor d-axis winding results in the following torque expression on the q-axis rotor winding:

$$T_{q,\text{rotor}} = -\frac{p}{2}\underbrace{(L_m i_{sd} + L_r i_{rd})}_{\lambda_{rd}} i_{rq} = -\frac{p}{2}\lambda_{rd} i_{rq}.$$

(3-44)

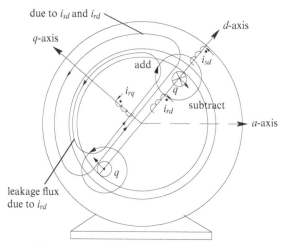

Fig. 3-9   Torque on the rotor *q*-axis.

3-5-3   Net Electromagnetic Torque T_{em} on the Rotor

By superposition, adding the torques acting on the *d*-axis and the *q*-axis of the rotor windings, the instantaneous torque is

$$T_{em} = T_{d,\text{rotor}} + T_{q,\text{rotor}}, \tag{3-45}$$

which, using Eq. (3-43) and Eq. (3-44), results in

$$T_{em} = \frac{p}{2}(\lambda_{rq}i_{rd} - \lambda_{rd}i_{rq}). \tag{3-46}$$

Substituting for flux linkages in Eq. (3-46), the electromagnetic torque can be expressed in terms of inductances as

$$T_{em} = \frac{p}{2}L_m(i_{sq}i_{rd} - i_{sd}i_{rq}). \tag{3-47}$$

3-6   ELECTRODYNAMICS

The acceleration is determined by the difference of the electromagnetic torque and the load torque (including friction torque) acting on J_{eq}, the combined inertia of the load and the motor. In terms of the actual (mechanical) speed of the rotor ω_{mech} in radians per second, where $\omega_{\text{mech}} = (2/p)\omega_m,$

$$\frac{d}{dt}\omega_{mech} = \frac{T_{em} - T_L}{J_{eq}}. \tag{3-48}$$

3-7   *d-* AND *q*-AXIS EQUIVALENT CIRCUITS

Substituting for flux linkage derivatives in terms of inductances into the voltage equations (Eq. 3-28 and Eq. 3-29 for the stator and Eq. 3-31 and Eq. 3-32 for the rotor),

$$v_{sd} = R_s i_{sd} - \omega_d \lambda_{sq} + L_{\ell s} \frac{d}{dt} i_{sd} + L_m \frac{d}{dt}(i_{sd} + i_{rd}) \tag{3-49}$$

$$v_{sq} = R_s i_{sq} + \omega_d \lambda_{sd} + L_{\ell s} \frac{d}{dt} i_{sq} + L_m \frac{d}{dt}(i_{sq} + i_{rq}) \tag{3-50}$$

and

$$\underset{=0}{\underline{v_{rd}}} = R_r i_{rd} - \omega_{dA} \lambda_{rq} + L_{\ell r} \frac{d}{dt} i_{rd} + L_m \frac{d}{dt}(i_{sd} + i_{rd}) \tag{3-51}$$

$$\underset{=0}{\underline{v_{rq}}} = R_r i_{rq} + \omega_{dA} \lambda_{rd} + L_{\ell r} \frac{d}{dt} i_{rq} + L_m \frac{d}{dt}(i_{sq} + i_{rq}). \tag{3-52}$$

For each axis, the stator and the rotor winding equations are combined to result in the *dq* equivalent circuits shown in Fig. 3-10a,b. Using Eq. (3-28), we can label the terminals across which the voltage is $d\lambda_{sd}/dt$

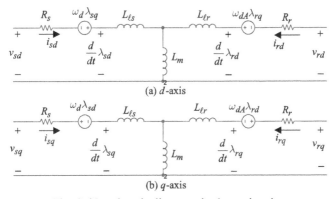

Fig. 3-10   *dq*-winding equivalent circuits.

in Fig. 3-10a. Similarly, using Eq. (3-29), Eq. (3-31), and Eq. (3-32), respectively, we can label terminals in Fig. 3-10a,b with $d\lambda_{sq}/dt$, $d\lambda_{rd}/dt$, and $d\lambda_{rq}/dt$.

3-8   RELATIONSHIP BETWEEN THE dq WINDINGS AND THE PER-PHASE PHASOR-DOMAIN EQUIVALENT CIRCUIT IN BALANCED SINUSOIDAL STEADY STATE

In this section, we will see that under a balanced sinusoidal steady-state condition, the dq-winding equations combine to result in the per-phase equivalent circuit of an induction machine that we have derived in the previous course [1]. It will be easiest to choose $w_d = w_{syn}$ (although any other choice of reference speed would lead to the same results; see Problem 3-8) so that the dq-winding quantities are dc and their time derivatives are zero under a balanced sinusoidal steady-state condition. Therefore, in the stator voltage equation Eq. (3-27) in steady state

$$\vec{v}_{s\_dq} = R_s \vec{i}_{s\_dq} + jw_{syn}\vec{\lambda}_{s\_dq} \quad \text{(steady state).} \qquad (3\text{-}53)$$

Similarly, the voltage equation for the rotor dq windings under a balanced sinusoidal steady state with $w_d = w_{syn}$ (thus, $w_{dA} = w_{slip} = sw_{syn}$) results in

$$0 = \frac{R_r}{s}\vec{i}_{r\_dq} + jw_{syn}\vec{\lambda}_{r\_dq} \quad \text{(steady state),} \qquad (3\text{-}54)$$

where s is the slip. Substituting for flux linkage space vectors in Eq. (3-53) and Eq. (3-54) results in

$$\vec{v}_{s\_dq} = R_s\vec{i}_{s\_dq} + jw_{syn}L_{\ell s}\vec{i}_{s\_dq} + jw_{syn}L_m(\vec{i}_{s\_dq} + \vec{i}_{r\_dq}) \qquad (3\text{-}55)$$

and

$$0 = \frac{R_r}{s}\vec{i}_{r\_dq} + jw_{syn}L_{\ell r}\vec{i}_{r\_dq} + jw_{syn}L_m(\vec{i}_{s\_dq} + \vec{i}_{r\_dq}). \qquad (3\text{-}56)$$

The above space vector equations in a balanced sinusoidal steady state correspond to the following phasor equations for phase a:

$$\bar{V}_a = R_s\bar{I}_a + jw_{syn}L_{\ell s}\bar{I}_a + jw_{syn}L_m(\bar{I}_a + \bar{I}_A) \qquad (3\text{-}57)$$

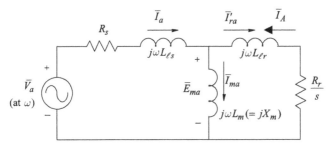

Fig. 3-11   Per-phase equivalent circuit in steady state.

and

$$0 = \frac{R_r}{s}\bar{I}_A + j\omega_{syn}L_{\ell r}\bar{I}_A + j\omega_{syn}L_m(\bar{I}_a + \bar{I}_A). \qquad (3\text{-}58)$$

The above two equations combined correspond to the per-phase equivalent circuit of Fig. 3-11 that was derived in the previous course [1] under a balanced sinusoidal steady-state condition. Note that in Fig. 3-11, $\bar{I}_A = -\bar{I}'_{ra}$.

3-9   COMPUTER SIMULATION

In *dq* windings, the flux linkages and voltage equations are derived earlier. We will use λ_{sd}, λ_{sq}, λ_{rd}, and λ_{rq} as state variables, and express i_{sd}, i_{sq}, i_{rd}, and i_{rq} in terms of these state variables. The reason for choosing flux linkages as state variables has to do with the fact that these quantities change slowly compared with currents, which can change *almost* instantaneously.

We can calculate *dq*-winding currents from the stator and the rotor flux linkages of the respective windings as follows: Referring to Fig. 3-3, the stator and the rotor *d*-winding flux linkages are related to their winding currents (rewriting Eq. 3-19 and Eq. 3-21 in a matrix form) as

$$\begin{bmatrix} \lambda_{sd} \\ \lambda_{rd} \end{bmatrix} = \underbrace{\begin{bmatrix} L_s & L_m \\ L_m & L_r \end{bmatrix}}_{[L]} \begin{bmatrix} i_{sd} \\ i_{rd} \end{bmatrix}. \qquad (3\text{-}59)$$

Similarly, in the q-axis windings, from Eq. (3-20) and Eq. (3-22), the matrix $[L]$ of the above equation relates flux linkages to respective currents

$$\begin{bmatrix} \lambda_{sq} \\ \lambda_{rq} \end{bmatrix} = \underbrace{\begin{bmatrix} L_s & L_m \\ L_m & L_r \end{bmatrix}}_{[L]} \begin{bmatrix} i_{sq} \\ i_{rq} \end{bmatrix}. \tag{3-60}$$

Combining matrix Eq. (3-59) and Eq. (3-60), we can relate fluxes to currents as follows:

$$\begin{bmatrix} \lambda_{sd} \\ \lambda_{sq} \\ \lambda_{rd} \\ \lambda_{rq} \end{bmatrix} = \underbrace{\begin{bmatrix} L_s & 0 & L_m & 0 \\ 0 & L_s & 0 & L_m \\ L_m & 0 & L_r & 0 \\ 0 & L_m & 0 & L_r \end{bmatrix}}_{[M]} \begin{bmatrix} i_{sd} \\ i_{sq} \\ i_{rd} \\ i_{rq} \end{bmatrix}. \tag{3-61}$$

From Eq. (3-61), currents can be calculated by using the inverse of matrix [M]:

$$\begin{bmatrix} i_{sd} \\ i_{sq} \\ i_{rd} \\ i_{rq} \end{bmatrix} = [M]^{-1} \begin{bmatrix} \lambda_{sd} \\ \lambda_{sq} \\ \lambda_{rd} \\ \lambda_{rq} \end{bmatrix}. \tag{3-62}$$

With voltages as input, and choosing the speed ω_d of the dq windings, the flux linkages are calculated as shown in Fig. 3-7 using Eq. (3-38) and Eq. (3-39) derived earlier. The currents are calculated using Eq. (3-62) derived above. Combining these with the torque equation Eq. (3-47) and the electrodynamics Eq. (3-48), we can draw the overall block diagram for computer modeling in Fig. 3-12.

3-9-1   Calculation of Initial Conditions

In order to carry out computer simulations, we need to calculate initial values of the state variables, that is, of the flux linkages of the dq windings. These can be calculated in terms of the initial values of the dq winding currents. These currents allow us to compute the electromagnetic torque in steady state, thus the initial loading of the induction

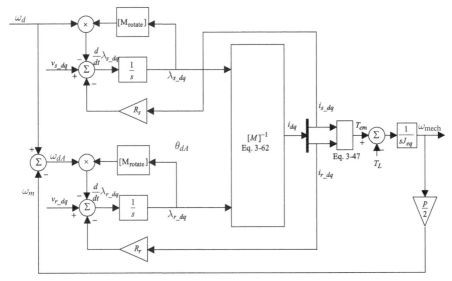

Fig. 3-12   Induction motor model in terms of *dq* windings.

machine. To accomplish this, we will make use of the phasor analysis in the initial steady state as follows:

Phasor Analysis

In the sinusoidal steady state, we can calculate current phasors $\bar{I}_a$ and $\bar{I}'_{ra} (= -\bar{I}_A)$ in Fig. 3-11 for a given $\bar{V}_a$. All the space vectors and the *dq*-winding variables at $t = 0$ can be calculated. The phasor current for phase-*a* allows the stator current space vector at $t = 0$ to be calculated as follows:

$$\bar{I}_a = \hat{I}_a \angle \theta_i \Rightarrow \underbrace{\vec{i}_s(0) = \frac{3}{2}\hat{I}_a\, e^{j\theta_i}}_{i_s}. \qquad (3\text{-}63)$$

Assuming the initial value of θ_{da} to be zero (i.e., the *d*-axis along the stator *a*-axis), and using Fig. 3-3 and Eq. (3-5) and Eq. (3-6)

$$i_{sd}(0) = \sqrt{\frac{2}{3}} \times \text{projection of } \vec{i}_s(0) \text{ on } d\text{-axis} = \sqrt{\frac{2}{3}}\underbrace{\left(\frac{3}{2}\hat{I}\right)}_{\hat{i}_s}\cos(\theta_i) \qquad (3\text{-}64)$$

and

$$i_{sq}(0) = \sqrt{\frac{2}{3}} \times \text{projection of } \vec{i}_s(0) \text{ on } q\text{-axis} = \sqrt{\frac{2}{3}} \underbrace{\left(\frac{3}{2}\hat{I}\right)}_{\hat{i}_s} \sin(\theta_i). \quad (3\text{-}65)$$

Similarly, we can calculate $\nu_{sd}(0)$ and $\nu_{sq}(0)$. The phasor $\bar{I}_A \, (= -\bar{I}'_{ra})$ allows $i_{rd}(0)$ and $i_r (0)$ to be calculated. Knowing the currents in the *dq* windings at $t = 0$ allows the initial values of the flux linkages to be calculated from Eq. (3-61). Also, these currents allow the computation of the electromagnetic torque in steady state by Eq. (3-47) to calculate the initial loading of the induction machine.

EXAMPLE 3-1

An induction machine with the following nameplate data is initially operating under its rated condition in steady state, supplying its rated torque. Calculate initial values of flux linkages and the load torque. Assume that the phase-*a* voltage has its positive peak at time $t = 0$.

Nameplate Data

Power: 3 HP/2.4 kW
Voltage: 460V (L-L, rms)
Frequency: 60 Hz
Phases: 3
Full-Load Current: 4 A
Full-Load Speed: 1750 rpm
Full-Load Efficiency: 88.5%
Power Factor: 80.0%
Number of Poles: 4

Per-Phase Motor Circuit Parameters

$R_s = 1.77 \, \Omega$

$R_r = 1.34 \, \Omega$

$X_{\ell s} = 5.25 \, \Omega \, (\text{at } 60 \, \text{Hz})$

$X_{\ell r} = 4.57 \, \Omega \, (\text{at } 60 \, \text{Hz})$

$X_m = 139.0 \, \Omega \, (\text{at } 60 \, \text{Hz})$

Full-Load Slip $= 1.72\%$

The iron losses are specified as 78 W and the mechanical (friction and windage) losses are specified as 24 W. The inertia of the machine is given. Assuming that the reflected load inertia is approximately the same as the motor inertia, the total equivalent inertia of the system is $J_{eq} = 0.025 \, \text{kg} \cdot \text{m}^2$.

Solution

A MATLAB file EX3_1.m on accompanying website is based on the following steps:

Step 1   Calculate by phasor analysis $\bar{V}_a$, $\bar{I}_a$, and $\bar{I}_A \, (= -\bar{I}'_{ra})$, given that the phase-*a* voltage has a positive peak at time $t = 0$.

Step 2   Calculate the current space vectors $\vec{i}_s^{\,a}$ and $\vec{i}_r^{\,a}$ at time $t = 0$ from the phasors for phase-*a*.

Step 3   In the *dq* analysis, choose $\omega_d = \omega_{syn}$ and $\theta_{da}(0) = 0$. Calculate $i_{s\alpha}$, $i_{s\beta}$, $i_{r\alpha}$, and $i_{r\beta}$ from the space vectors in step 2, using equations similar to Eq. (3-64) and Eq. (3-65).

Step 4   Calculate flux linkages of *dq* windings using Eq. (3-61).

Step 5   Calculate torque $T_L(0)$, which equals T_{em} in steady state, from Eq. (3-46) or Eq. (3-47).

The results from EX3_1.m are listed below:

$\lambda_{sd}(0) = 0.0174$ Wb-turns

$\lambda_{rd}(0) = -0.1237$ Wb-turns

$\lambda_{sq}(0) = -1.1951$ Wb-turns

$\lambda_{rq}(0) = -1.1363$ Wb-turns

$T_{em}(0) = T_L(0) = 12.644$ Nm.

EXAMPLE 3-2

Calculate the initial conditions of the induction machine operating in steady state in Example 3-1 using the voltage equations Eq. (3-28), Eq. (3-29), Eq. (3-31), and Eq. (3-32). Also calculate the load torque.

Solution

In a balanced steady state, with ω_d chosen as the synchronous speed ω_{syn}, all *dq* winding variables are dc quantities and $\omega_{dA} = s\omega_{syn}$. Therefore, their time derivatives are zero in Eq. (3-28), Eq. (3-29), Eq. (3-31), and Eq. (3-32), resulting in the following equations:

$$v_{sd} = R_s i_{sd} - \omega_{syn}\lambda_{sq} \tag{3-66}$$

$$v_{sq} = R_s i_{sq} + \omega_{syn}\lambda_{sd} \tag{3-67}$$

$$0 = R_r i_{rd} - s\omega_{syn}\lambda_{rq} \tag{3-68}$$

$$0 = R_r i_{rq} + s\omega_{syn}\lambda_{rd}. \tag{3-69}$$

Substituting in the above equations for flux linkages from Eq. (3-19) through Eq. (3-22)

$$\begin{bmatrix} v_{sd} \\ v_{sq} \\ 0 \\ 0 \end{bmatrix} = \underbrace{\begin{bmatrix} R_s & -\omega_{syn}L_s & 0 & -\omega_{syn}L_m \\ \omega_{syn}L_s & R_s & \omega_{syn}L_m & 0 \\ 0 & -s\omega_{syn}L_m & R_r & -s\omega_{syn}L_r \\ s\omega_{syn}L_m & 0 & s\omega_{syn}L_r & R_r \end{bmatrix}}_{[A]} \begin{bmatrix} i_{sd} \\ i_{sq} \\ i_{rd} \\ i_{rq} \end{bmatrix}. \tag{3-70}$$

The machine currents can be calculated from Eq. (3-70) by inverting matrix $[A]$:

$$\begin{bmatrix} i_{sd} \\ i_{sq} \\ i_{rd} \\ i_{rq} \end{bmatrix} = [A]^{-1} \begin{bmatrix} v_{sd} \\ v_{sq} \\ 0 \\ 0 \end{bmatrix}. \tag{3-71}$$

Once the *dq* winding currents are calculated, the flux linkages can be calculated from Eq. (3-61).

The results from a MATLAB file EX3_2.m on the accompanying website are as follows for currents, with flux linkages and the load torque exactly as in Example 3-1.

$i_{sd}(0) = 5.34$ A

$i_{sq}(0) = -3.7$ A

$i_{rd}(0) = -5.5$ A

$i_{rq}(0) = 0.60$ A.

EXAMPLE 3-3

In Simulink, develop a simulation of the induction machine described in Example 3-1 operating in steady state as specified in Example 3-1. At $t = 0.1$ seconds, the load torque T_L suddenly goes to one-half of its initial value and stays at that level. Assume $\theta_{da}(0) = 0$ and a synchronously rotating dq reference frame.

Plot the electromagnetic torque developed by the motor and the rotor speed as functions of time.

Solution

The Simulink file EX3_3.mdl included on the accompanying website follows the block diagram in Fig. 3-12. Prior to its execution, initial conditions for the flux linkages, the rotor speed, and the load torque must be calculated either by executing the file for Example 3-1 (EX3_1.m) or for Example 3-2 (EX3_2.m) by double clicking on the start icon shown in the schematic of Fig. 3-13. The resulting waveforms are plotted in Fig. 3-14.

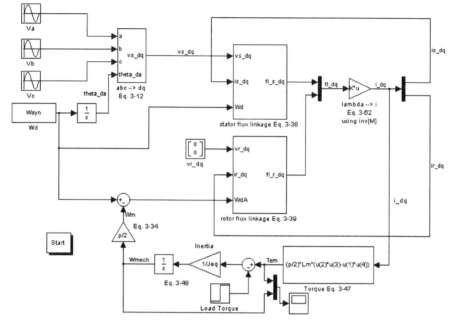

Fig. 3-13   Simulation of Example 3-3.

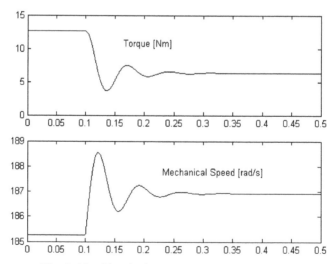

Fig. 3-14   Simulation results of Example 3-3.

EXAMPLE 3-4

Consider the "test" machine described in Chapter 1. Simulate the machine starting from a completely powered-down state with no external load connected to it. Simulate for three different assumptions regarding the d-axis:

(a) $\omega_d = \omega_{syn}$
(b) $\omega_d = \omega_m$
(c) The d-axis is aligned with the rotor flux linkage space vector.

Solution

See the complete solution on the accompanying website.

EXAMPLE 3-5

Consider the "test" machine described in Chapter 1. Assume that this machine is operating under its rated condition in steady state, supplying its rated torque. Calculate the initial values of the flux linkages of dq windings, the dq currents, the rotor speed, and the torque for the following three assumptions:

(a) $\omega_d = \omega_{syn}$; assume $\theta_{da}(0) = 0$ and $\theta_{dA}(0) = 0$.
(b) $\omega_d = \omega_m$; assume $\theta_{da}(0) = 0$ and $\theta_{dA}(0) = 0$.
(c) The d-axis is aligned with the rotor flux linkage space vector.

Solution

See the complete solution on the accompanying website.

EXAMPLE 3-6

Use the initial conditions calculated in Example 3-5 to simulate a sudden change in load torque to one-half of its initial value. Plot various quantities as a function of time.

Solution

See the complete solution on the accompanying website.

EXAMPLE 3-7

The machine in Example 3-5 is made to go into the generator mode. Make the load torque change linearly from its initial rated positive value to the rated negative value in 0.2 seconds. Plot various quantities as a function of time.

Solution

See the complete solution on the accompanying website.

3-10 SUMMARY

In this chapter, the actual stator phase windings and the equivalent rotor phase windings are represented by an equivalent set of *dq* windings, which produce the same air gap mmf. There are many advantages of doing so: there is zero magnetic coupling between the windings on the *d*-axis and those on the *q*-axis due to their orthogonal orientation. This procedure results in much simpler expressions and allows air gap flux and electromagnetic torque to be controlled independently, which will be discussed in the following chapters.

REFERENCE

1. N. Mohan, *Electric Machines and Drives*, Wiley, Hoboken, NJ, 2012. http://www.wiley.com/college/mohan.

PROBLEMS

3-1 Derive that each of the two windings with $\sqrt{3/2}N_s$ turns in an equivalent two-phase machine has the magnetizing inductance of L_m, leakage inductance of $L_{\ell s}$, and the resistance R_s, where these quantities correspond to those of an equivalent three-phase machine.

3-2 Show that the instantaneous power loss in the stator resistances is the same in the three-phase machine as in an equivalent two-phase machine.

3-3 Show that the instantaneous total input power is the same in a-b-c and in the dq circuits.

3-4 Rederive all the equations used in obtaining the torque expressions of Eq. (3-46) and Eq. (3-47) for a p-pole machine from energy considerations.

3-5 Draw dynamic equivalent circuits of Fig. 3-9 for the following values of the dq winding speed ω_d: 0 and ω_m. What is the frequency of dq winding variables in a balanced sinusoidal steady state, including the condition that $\omega_d = \omega_{syn}$?

3-6 The "test" motor described in Chapter 1 is operating at its rated conditions. Calculate $\nu_{sd}(t)$ and $\nu_{sq}(t)$ as functions of time, (a) if $\omega_d = \omega_{syn}$, and (b) if $\omega_d = 0$.

3-7 Under a balanced sinusoidal steady state, calculate the input power factor of operation based on the d- and the q-axis equivalent circuits of Fig. 3-10 in the "test" motor described in Chapter 1.

3-8 Show that the equations for the dq windings in a balanced sinusoidal steady state result in the per-phase equivalent circuit of Fig. 3-11 for $\omega_d = 0$ and $\omega_d = \omega_m$.

3-9 Using Eq. (3-46) and Eq. (3-47), derive the torque expressions in terms of (1) stator dq winding flux linkages and currents, and (2) stator dq winding currents and rotor dq winding flux linkages.

3-10 Show that for the transformation matrix in Eq. (3-12),

$$\underbrace{[T]_{abc\to dq}}_{2\times 3}\underbrace{[T]_{dq\to abc}}_{3\times 2}=\underbrace{[I]}_{2\times 2}.$$

3-11 Derive the voltage equations in the *dq* stator windings (Eq. 3-28 and Eq. 3-29) using the transformation matrix of Eq. (3-12) (also for the rotor *dq* windings).

3-12 In Examples 3-1 and 3-2, plot $\vec{v}_s$, $\vec{\lambda}_s$, $\vec{\lambda}_r$, $\vec{i}_s$, and $\vec{i}_r$ at time $t = 0$.

3-13 In the simulation of Example 3-3, plot the *dq* winding currents and the *a-b-c* phase currents. Also plot the slip speed.

3-14 Repeat Example 3-3, assuming $\omega_d = 0$. Plot the *dq* winding currents, the *a-b-c* phase currents, and the slip speed.

3-15 Repeat Example 3-3 assuming $\omega_d = \omega_m$. Plot the *dq* winding currents, the *a-b-c* phase currents, and the slip speed.

3-16 Modify the simulation file of Example 3-3 to simulate line start with the rated load connected to the motor, without the load disturbance at $t = 0.1$ seconds.

3-17 The "test" machine is made to go into the generator mode. Modify the file of Example 3-3 by making the load toque change linearly from its initial rated positive value to the rated negative value in 0.2 seconds, starting at $t = 0.1$ seconds. Plot the same variables as in that example, as well as the phase voltages and currents.

4 Vector Control of Induction-Motor Drives: A Qualitative Examination

4-1 INTRODUCTION

Applications such as robotics and factory automation require accurate control of speed and position. This can be accomplished by vector control of induction machines, which emulate the performance of dc motor and brushless dc motor servo drives. Compared with dc and brushless dc motors, induction motors have a lower cost and a more rugged construction.

In any speed and position control application, torque is the fundamental variable that needs to be controlled. The ability to produce a step change in torque on command represents total control over the drives for high performance speed and position control.

This chapter qualitatively shows how a step change in torque is accomplished by vector control of induction-motor drives. For this purpose, the steady-state analysis of induction motors discussed in the previous course serves very well because while delivering a step change in electromagnetic torque under vector control, an induction machine instantaneously transitions from one steady state to another.

4-2 EMULATION OF dc AND BRUSHLESS dc DRIVE PERFORMANCE

Under vector control, induction-motor drives can emulate the performance of dc-motor and brushless-dc motor servo drives discussed

Advanced Electric Drives: Analysis, Control, and Modeling Using MATLAB/Simulink®, First Edition. Ned Mohan.

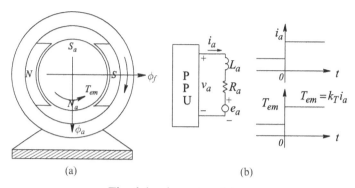

Fig. 4-1   dc motor drive.

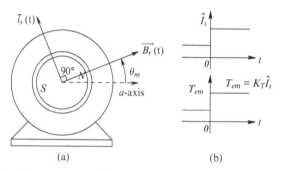

Fig. 4-2   Current-controlled brushless dc (BLDC) motor drive.

in the previous course. These are briefly reviewed in the following sections.

In the dc-motor drive shown in Fig. 4-1a, the commutator and brushes ensure that the armature-current-produced magnetomotive force (mmf) is at a right angle to the field flux produced by the stator. Both of these fields remain stationary. The electromagnetic torque T_{em} developed by the motor depends linearly on the armature current i_a:

$$T_{em} = k_T i_a, \tag{4-1}$$

where k_T is the dc motor torque constant. To change T_{em} as a step, the armature current i_a is changed (at least, attempted to be changed) as a step by the power-processing unit (PPU), as shown in Fig. 4-1b.

In the brushless-dc drive shown in Fig. 4-2a, the PPU keeps the stator current space vector $\vec{i}_s(t)$ 90° ahead of the rotor field vector $\vec{B}_r(t)$

(produced by the permanent magnets on the rotor) in the direction of rotation. The position $\theta_m(t)$ of the rotor field is measured by means of a sensor, for example, a resolver. The torque T_{em} depends on $\hat{I}_s$, the amplitude of the stator current space vector $\vec{i}_s(t)$:

$$T_{em} = k_T \hat{I}_s, \qquad (4\text{-}2)$$

where k_T is the brushless dc motor torque constant. To produce a step change in torque, the PPU changes the amplitude $\hat{I}_s$ in Fig. 4-2b by appropriately changing $i_a(t)$, $i_b(t)$, and $i_c(t)$, keeping $\vec{i}_s(t)$ always ahead of $\vec{B}_r(t)$ by 90° in the direction of rotation.

4-2-1   Vector Control of Induction-Motor Drives

We will look at one of many ways in which an induction motor drive can emulate the performance of dc and brushless dc motor drives. Based on the steady state analysis in Reference [1], we observed that in an induction machine, $\vec{F}_r(t)$ and $\vec{F}_r'(t)$ space vectors are naturally at 90° to the rotor flux-density space vector $\vec{B}_r(t)$, as shown in Fig. 4-3a. In terms of the amplitude $\hat{I}_r'$, where

$$\vec{i}_r'(t) = \frac{\vec{F}_r'(t)}{N_s / p},$$

keeping $\hat{B}_r$ constant results in the following torque expression:

$$T_{em} = k_T \hat{I}_r', \qquad (4\text{-}3)$$

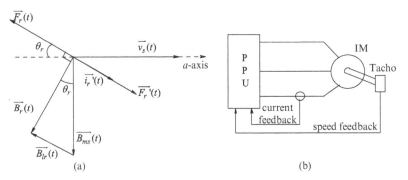

(a)   (b)

Fig. 4-3   (a) Rotor flux density and mmf space vectors; (b) vector-controlled induction-motor drive.

where k_T is the induction-motor torque constant. The earlier discussion shows that induction-motor drives can emulate the performance of dc motor and the brushless dc motor drives. In induction machines, in this emulation (called vector control), the PPU in Fig. 4-3b controls the stator current space vector $\vec{i}_s(t)$ as follows: a component of $\vec{i}_s(t)$ is controlled to keep $\hat{B}_r$ constant, while the other orthogonal component of $\vec{i}_s(t)$ is controlled to produce the desired torque.

4-3 ANALOGY TO A CURRENT-EXCITED TRANSFORMER WITH A SHORTED SECONDARY

To understand vector control in induction-motor drives, an analogy of a current-excited transformer with a short-circuited secondary, as shown in Fig. 4-4, is very useful. Initially at time $t = 0^-$, both currents and the core flux are zero. The primary winding is excited by a step-current at $t = 0^+$. Changing this current as a step, in the presence of leakage fluxes, requires a voltage impulse, but as has been argued in Reference [2], the volt-seconds needed to bring about such a change are not all that large. In any case, we will initially assume that it is possible to produce a step change in the primary winding current. Our focus is on the short-circuited secondary winding; therefore, we will neglect the leakage impedance of the primary winding.

In the transformer of Fig. 4-4 at $t = 0^-$, the flux linkage $\lambda_2(0^-)$ of the secondary winding is zero, as there is no flux in the core. From the theorem of constant flux linkage, we know that the flux linkage of a short-circuited coil cannot change instantaneously. Therefore, at $t = 0^+$,

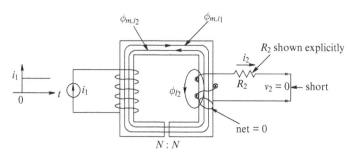

Fig. 4-4 Current-excited transformer with secondary short-circuited.

$$\lambda_2(0^+) = \lambda_2(0^-) = 0. \tag{4-4}$$

To maintain the above condition, i_2 will jump instantaneously at $t = 0^+$. As shown in Fig. 4-4 at $t = 0^+$, there are three flux components linking the secondary winding: the magnetizing flux ϕ_{m,i_1} produced by i_1, the magnetizing flux ϕ_{m,i_2} produced by i_2, and the leakage flux $\phi_{\ell 2}$ produced by i_2, which links only winding 2 but not winding 1. The condition that $\lambda_2(0^+) = 0$ requires that the net flux linking winding 2 be zero; hence, including the flux directions shown in Fig. 4-4,

$$\phi_{m,i_1}(0^+) - \phi_{m,i_2}(0^+) - \phi_{\ell 2}(0^+) = 0$$

or

$$\phi_{m,i_2}(0^+) + \phi_{\ell 2}(0^+) = \phi_{m,i_1}(0^+). \tag{4-5}$$

Choosing the positive flux direction to be in the downward direction through coil 2, the flux linkage of coil 2 can be written as

$$\lambda_2 = -\underbrace{N\phi_{m,i_2}}_{L_m i_2} - \underbrace{N\phi_{\ell 2}}_{L_{\ell 2} i_2} + \underbrace{N\phi_{m,i_1}}_{L_m i_1} \tag{4-6a}$$

or

$$\lambda_2 = -L_2 i_2 + L_m i_1, \tag{4-6b}$$

where

$$L_m = \text{the mutual inductance between the two coils} \tag{4-7}$$

and

$$L_2 = L_m + L_{\ell 2} = \text{the self-inductance of coil 2.} \tag{4-8}$$

Therefore at $t = 0^+$,

$$\lambda_2(0^+) = 0 = -L_2 i_2(0^+) + L_m i_1(0^+) \tag{4-9}$$

or

$$i_2(0^+) = \frac{L_m}{L_2} i_1(0^+). \tag{4-10}$$

We should note that the secondary winding current $i_2(0^+)$ does not depend on the secondary winding resistance R_2. Also, if the secondary

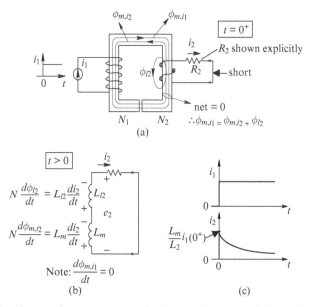

Fig. 4-5 Analogy of a current-excited transformer with a short-circuited secondary; $N_1 = N_2$.

winding leakage flux is neglected ($L_{\ell 2} = 0$), then $i_2(0^+)$ equals $i_1(0^+)$, with the assumption of unity turns-ratio between the two windings.

In the current-excited transformer of Fig. 4-5a, the equivalent circuit of the secondary winding for $t > 0$ is shown in Fig. 4-5b. Note that in Fig. 4-5a for $t > 0$, ϕ_{m,i_1} is a constant and therefore does not induce any voltage in Fig. 4-5b. Due to the voltage drop across R_2, the current i_2 declines, thus causing both ϕ_{m,i_2} and $\phi_{\ell 2}$ to decline (both of these are produced by i_2). In accordance with the equivalent circuit of Fig. 4-5b, i_2 decays exponentially, as shown in Fig. 4-5c

$$i_2(t) = i_2(0^+)e^{-t/\tau_2}, \tag{4-11}$$

where τ_2 is the time constant of winding 2:

$$\tau_2 = \frac{L_2}{R_2}. \tag{4-12}$$

Theoretically, we can see that at $t = 0^+$, a voltage impulse is necessary to make the current i_1 jump because a finite amount of energy must be transferred instantaneously. This instantaneous energy increase (all of these associated energy levels were zero at $t = 0^-$) is associated with:

1. Leakage flux of the primary (neglected in this discussion)
2. Leakage flux of the secondary in air
3. Slight increase of flux $\phi_{\ell 2}(= \phi_{m,i_1} - \phi_{m,i_2})$ in the core.

However, as argued in Reference [2], the volt-seconds needed to accomplish this instantaneous change in current are not excessive. After all, note that in a dc-motor drive, for a step change in torque, the armature current must be built up overcoming the inductive nature of the armature winding. A similar situation occurs in "brushless-dc" motor drives.

4-3-1 Using the Transformer Equivalent Circuit

The earlier discussion can also be confirmed by considering the equivalent circuit of a two winding transformer with $N_1 = N_2$, shown in Fig. 4-6. For a step change in i_1 at $t = 0^+$, the instantaneous current division is based on inductances of the two parallel branches (resistance R_2 will have a negligible effect):

$$i_2(0^+) = \frac{L_m}{L_m + L_{\ell 2}} i_1(0^+) = \frac{L_m}{L_2} i_1(0^+) \tag{4-13}$$

and

$$i_m(0^+) = \frac{L_{\ell 2}}{L_m + L_{\ell 2}} i_1(0^+) = \frac{L_{\ell 2}}{L_2} i_1(0^+). \tag{4-14}$$

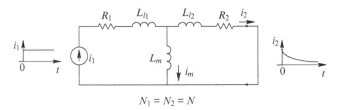

$$N_1 = N_2 = N$$

Fig. 4-6 Equivalent-circuit representation of the current-excited transformer with a short-circuited secondary.

Equation (4-13) shows the jump in i_2 at $t = 0^+$. Noting that $di_1/dt = 0$ for $t > 0$, solving for i_2 in the circuit in Fig. 4-6 confirms the decay in i_2 due to R_2

$$i_2(t) = i_2(0^+)e^{-t/\tau_2}. \tag{4-15}$$

4-4   d- AND q-AXIS WINDING REPRESENTATION

A step change in torque requires a step change in the rotor current of a vector-controlled induction motor. We will make use of an orthogonal set of d- and q-axis windings, introduced in Chapter 3, producing the same mmf as three stator windings (each with N_s turns, sinusoidally distributed), with i_a, i_b, and i_c flowing through them. In Fig. 4-7 at a time t, $\vec{i}_s(t)$ and $\vec{F}_s(t)$ are produced by $i_a(t)$, $i_b(t)$, and $i_c(t)$. The resulting mmf $\vec{F}_s(t) = (N_s / p)\vec{i}_s(t)$ can be produced by the set of orthogonal stator windings shown in Fig. 4-7, each sinusoidally distributed with $\sqrt{3/2}N_s$ turns: one winding along the d-axis, and the other along the q-axis. Note that this d–q axis set may be at any arbitrary angle with respect to the phase-a axis. In order to keep the mmf and the flux-density distributions the same as in the actual machine with three-phase windings, the currents in these two windings would have to be i_{sd} and i_{sq}, where, as shown

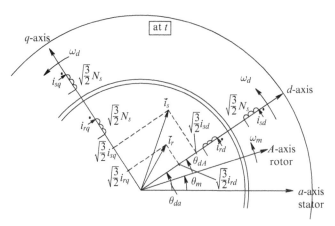

Fig. 4-7 Stator and rotor mmf representation by equivalent dq winding currents.

in Fig. 4-7, these two current components are $\sqrt{2/3}$ times the projections of the $\vec{i}_s(t)$ vector along the *d*-axis and *q*-axis.

4-5 VECTOR CONTROL WITH *d*-AXIS ALIGNED WITH THE ROTOR FLUX

In the following analysis, we will assume that the *d*-axis is always aligned with the rotor flux-linkage space vector, that is, also aligned with $\vec{B}_r(t)$.

4-5-1 Initial Flux Buildup Prior to $t = 0^-$

We will apply the information of the last section to vector control of induction machines. As shown in Fig. 4-8, prior to $t = 0^-$, the magnetizing currents are built up in three phases such that

$$i_a(0^-) = \hat{I}_{m,\text{rated}} \quad \text{and} \quad i_b(0^-) = i_c(0^-) = -\frac{1}{2}\hat{I}_{m,\text{rated}}. \qquad (4\text{-}16)$$

The current buildup prior to $t = 0^-$ may occur slowly over a long period of time and represents the buildup of the flux in the induction machine up to its rated value. These currents represent the rated magnetizing currents to bring the air gap flux density to its rated value. Note

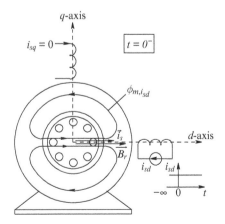

Fig. 4-8 Currents and flux at t^-.

that there will be no rotor currents at $t = 0^-$ (they decay out prior to $t = 0^-$). Also, at $t = 0^-$, the stator mmf can be represented by that produced by the d-axis winding (chosen to be along the a-axis) with a current i_{sd}, where

$$i_{sd}(0^-) = \sqrt{\frac{2}{3}}\hat{I}_{ms,\text{rated}} = \sqrt{\frac{2}{3}}\left(\frac{3}{2}\hat{I}_{m,\text{rated}}\right) = \sqrt{\frac{3}{2}}\hat{I}_{m,\text{rated}} \qquad (4\text{-}17)$$

and

$$i_{sq} = 0. \qquad (4\text{-}18)$$

We should note that the i_{sd}-produced stator leakage flux does not link the rotor, and hence it is of no concern in the following discussion.

At $t = 0^-$, the peak of the flux lines $\phi_{m,i_{sd}}$ linking the rotor is horizontally oriented. There is no rotor leakage flux because there are no currents flowing through the rotor bars. Only the flux $\phi_{m,i_{sd}}$ produced by the stator links the rotor. Therefore, $\vec{B}_r(0^-)$, equal to $\vec{B}_{ms}(0^-)$, is horizontally oriented along the d-axis (same as the a-axis at $t = 0^-$).

4-5-2 Step Change in Torque at $t = 0^+$

Next, we will see how this induction machine can produce a step change in torque. Initially, we will assume that the rotor is blocked from turning $(\omega_{mech} = 0)$, a restriction that will soon be removed. At $t = 0^+$, the three stator currents are changed as a step in order to produce a step change in the q-axis current i_{sq}, without changing i_{sd}, as shown in Fig. 4-9a. The current i_{sq} in the stator q winding produces the flux lines $\phi_{m,i_{sq}}$ that cross the air gap and link the rotor. The leakage flux produced by i_{sq} can be safely neglected from the discussion here (because it does not link the shorted rotor cage), similar to neglecting the leakage flux produced by the primary winding of the transformer in the previous analogy.

Turning our attention to the rotor at $t = 0^+$, we note that the rotor is a short-circuited cage, so its flux linkage cannot change instantaneously. To oppose the flux lines produced by i_{sq}, currents are instantaneously induced in the rotor bars by the transformer action, as shown in Fig. 4-9a.

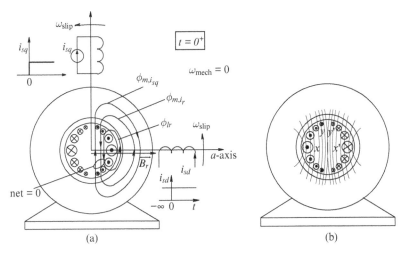

Fig. 4-9 Currents at $t = 0^+$.

This current distribution in the rotor bars is sinusoidal, as justified below using Fig. 4-9b:

> To justify the sinusoidal distribution of current in the rotor bars, assume that the bars $x - x'$ constitute one short-circuited coil, and the bars $y - y'$ the other coil. The density of flux lines produced by i_{sq} is sinusoidally distributed in the air gap. The coil $x - x'$ links most of the flux lines produced by i_{sq}. But the coil $y - y'$ links far fewer flux lines. Therefore, the current in this coil will be relatively smaller than the current in $x - x'$.

These rotor currents in Fig. 4-9a produce two flux components with peak densities along the *q*-axis and of the direction shown:

1. The magnetizing flux ϕ_{m,i_r} that crosses the air gap and links the stator.
2. The leakage flux $\phi_{\ell r}$ that does not cross the air gap and links only the rotor.

By the theorem of constant flux linkage, at $t = 0^+$, the net flux linking the short-circuited rotor in the *q*-axis must remain zero. Therefore, at $t = 0^+$, for the condition that $\phi_{rq,net} = 0$ (taking flux directions into account):

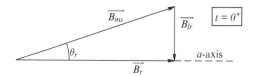

Fig. 4-10   Flux densities at $t = 0^+$.

$$\phi_{m,i_{sq}}(o^+) = \phi_{m,ir}(o^+) + \phi_{\ell r}(o^+). \qquad (4\text{-}19)$$

Since i_{sd} and the d-axis rotor flux linkage have not changed, the net flux, $\overrightarrow{B_r}$, linking the rotor remains the same at $t = 0^+$ as it was at $t = 0^-$.

The space vectors at $t = 0^+$ are shown in Fig. 4-10. No change in the net flux linking the rotor implies that $\overrightarrow{B_r}$ has not changed; its peak is still horizontal along the a-axis and of the same magnitude as before.

The rotor currents produced instantaneously by the transformer action at $t = 0^+$, as shown in Fig. 4-9a, result in a torque $T_{em}(0^+)$. This torque will be proportional to $\hat{B}_r$ and i_{sq} (slightly less than i_{sq} by a factor of L_m/L_r due to the rotor leakage flux, where L_r equals $L_m + L'_{\ell r}$ in the per-phase equivalent circuit of an induction machine):

$$T_{em} = k_1 \hat{B}_r \left(\frac{L_m}{L_r} i_{sq} \right), \qquad (4\text{-}20)$$

where k_1 is a constant. If no action is taken beyond $t = 0^+$, the rotor currents will decay and so will the force on the rotor bars. This current decay would be like in a transformer of Fig. 4-6 with a short-circuited secondary and with the primary excited with a step of current source.

In the transformer case of Fig. 4-6, decay of i_2 could be prevented by injecting a voltage equal to $R_2 i_2(0^+)$ beyond $t = 0^+$ to overcome the voltage drop across R_2. In the case of an induction machine, beyond $t = 0^+$, as shown in Fig. 4-11, we will equivalently rotate both the d-axis and the q-axis stator windings at an appropriate slip speed ω_{slip} in order to maintain $\overrightarrow{B_r}(t)$ completely along the d-axis with a constant amplitude of $\hat{B}_r$, and to maintain the same rotor-bar current distribution along the q-axis. This corresponds to the beginning of a new steady state. Therefore, the steady-state analysis of induction machine applies.

As the d-axis and the q-axis windings rotate at the appropriate value of ω_{slip} (notice that the rotor is still blocked from turning in Fig. 4-11),

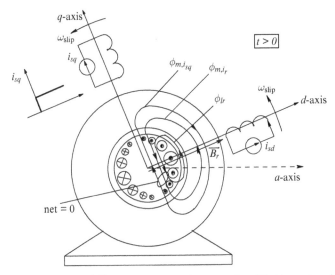

Fig. 4-11   Current and fluxes at some time $t > 0$, with the rotor blocked.

there is no net rotor flux linkage along the *q*-axis. The flux linkage along the *d*-axis remains constant with a flux density $\hat{B}_r$ "cutting" the rotor bars and inducing the bar voltages to cancel the iR_{bar} voltage drops. Therefore, the entire distribution rotates with time, as shown in Fig. 4-11 at any arbitrary time $t > 0$. For the relative distribution and hence the torque produced to remain the same as at $t = 0^+$, the two windings must rotate at an exact ω_{slip}, which depends linearly on both the rotor resistance R_r' and i_{sq} (slightly less by the factor L_m/L_r due to the rotor leakage flux), and inversely on $\hat{B}_r$

$$\omega_{slip} = k_2 \frac{R_r'(L_m / L_r)i_{sq}}{\hat{B}_r},\tag{4-21}$$

where k_2 is a constant. Now we can remove the restriction of $\omega_{mech} = 0$. If we need to produce a step change in torque while the rotor is turning at some speed ω_{mech}, then the *d*-axis and the *q*-axis windings should be equivalently rotated at the appropriate slip speed ω_{slip} relative to the rotor speed $\omega_m (= (p/2)\omega_{mech})$ in electrical rad/s, that is, at the synchronous speed $\omega_{yn} = \omega_m + \omega_{slip}$, as shown in Fig. 4-12.

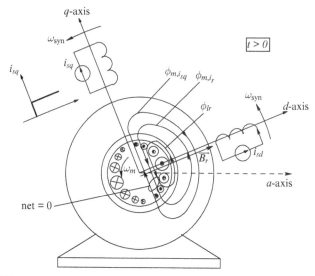

Fig. 4-12   Vector-controlled condition with the rotor speed $\omega_m\,(=(p/2)\omega_{\text{mech}})$ electrical rad/s.

4-6   TORQUE, SPEED, AND POSITION CONTROL

In vector control of induction-motor drives, the stator phase currents $i_a(t)$, $i_b(t)$, and $i_c(t)$ are controlled in such a manner that $i_{sq}(t)$ delivers the desired electromagnetic torque while $i_{sd}(t)$ maintains the peak rotor-flux density at its rated value. The reference values $i_{sq}^*(t)$ and $i_{sd}^*(t)$ are generated by the torque, speed, and position control loops, as discussed in the following section.

4-6-1   The Reference Current $i_{sq}^*(t)$

The reference value $i_{sq}^*(t)$ depends on the desired torque, which is calculated within the cascade control of Fig. 4-13, where the position loop is the outermost loop and the torque loop is the innermost loop. The loop bandwidths increase from the outermost to the innermost loop. The error between the reference (desired) position, $\theta_{\text{mech}}^*(t)$, and the measured position $\theta_{\text{mech}}(t)$ is amplified by a proportional (P) amplifier to generate the speed reference signal $\omega_{\text{mech}}^*(t)$. The error between the reference speed $\omega_{\text{mech}}^*(t)$ and the measured speed $\omega_{\text{mech}}(t)$ is amplified

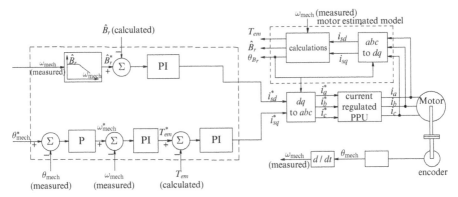

Fig. 4-13 Vector-controlled induction motor drive with a current-regulated PPU. In a multipole machine, measured speed should be converted into electrical radians per second.

by a proportional-integral (PI) amplifier to generate the torque reference $T_{em}^*(t)$. Finally, the error between $T_{em}^*(t)$ and the calculated torque $T_{em}(t)$ is amplified by another PI amplifier to generate the reference value $i_{sq}^*(t)$.

4-6-2   The Reference Current $i_{sd}^*(t)$

For measured speed values below the rated speed of the motor, the rotor flux-density peak $\hat{B}_r$ is maintained at its rated value as shown by the speed versus flux-density block in Fig. 4-13. Above the rated speed, the flux density is reduced in the flux-weakening mode, as discussed in the previous course. The error between $\hat{B}_r^*$ and the calculated flux-density peak $\hat{B}_r$ is amplified by a PI amplifier to generate the reference value $i_{sd}^*(t)$.

4-6-3   Transformation and Inverse-Transformation of Stator Currents

Fig. 4-13 shows the angle $\theta_{B_r}(t)$ of the d-axis, with respect to the stationary a-axis, to which the rotor flux-density space vector $\vec{B}_r(t)$ is aligned. The angle θ_{B_r} is the same as θ_{da} in Chapter 3 if the d-axis is aligned with the rotor flux $\vec{\lambda}_r$ at all times such that $\lambda_{rq} = 0$. This angle is computed by the vector-controlled motor model, which is described in the next section. Using the d-axis angle $\theta_{B_r}(t)$, the reference current signals $i^*_{sd}(t)$ and $i^*_{sq}(t)$ are transformed into the stator current reference signals $i^*_a(t)$, $i^*_b(t)$, and $i^*_c(t)$, as shown in Fig. 4-13 by the transform block (dq – to – abc), same as $[T_s]_{dq \to abc}$ in Equation (3-18) of the previous chapter. The current-regulated PPU uses these reference signals to supply the desired currents to the motor (details of how it can be accomplished are discussed briefly in Section 4-7).

The stator currents are measured and the d-axis angle $\theta_{B_r}(t)$ is used to transform them using a matrix same as $[T_s]_{abc \to dq}$ in Equation (3-12) of the previous chapter into the signals $i_{sd}(t)$ and $i_{sq}(t)$, as shown by the inverse transform block (abc – to – dq) in Fig. 4-13.

4-6-4   The Estimated Motor Model for Vector Control

The estimated motor model in Fig. 4-13 has the following measured inputs: the three stator phase currents $i_a(t)$, $i_b(t)$, and $i_c(t)$, and the measured rotor speed $\omega_{mech}(t)$. The motor model also needs accurate estimation of the rotor parameters L_m, $L'_{\ell r}$, and R'_r. The following parameters are calculated in the motor model for internal use and also as outputs: the angle θ_{B_r} (with respect to the stationary phase-a axis) to which the d-axis is aligned, the peak of the rotor flux density $\hat{B}_r(t)$, and the electromagnetic torque $T_{em}(t)$.

In the estimated motor model, $\hat{B}_r(t)$ is computed by considering the dynamics along the d-axis, which is valid in the flux-weakening mode, where $\hat{B}_r(t)$ is decreased to allow operation at higher than rated speed.

The electromagnetic torque $T_{em}(t)$ is computed based on Eq. (4-20) (the complete torque expression will be derived in the next chapter). The angle $\theta_{B_r}(t)$ is computed by first calculating the slip speed $\omega_{slip}(t)$ based on Eq. (4-21) (the complete expression will be derived in the next chapter). This slip speed is added to the measured rotor speed $\omega_m (= (p/2)\omega_{mech})$ to yield the instantaneous synchronous speed of the d- and the q-axes:

$$\omega_{syn}(t) = \omega_m(t) + \omega_{slip}(t). \tag{4-22}$$

With $\theta_{B_r} = 0$ at starting by initially aligning the rotor flux-density space vector along the a-axis, integrating the instantaneous synchronous speed results in the d-axis angle as follows:

$$\theta_{B_r}(t) = 0 + \int_0^t \omega_{\text{syn}}(\tau) \cdot d\tau, \tag{4-23}$$

where τ is a variable of integration. Based on these physical principles, the mathematical expressions are clearly and concisely developed in the next chapter.

4-7 THE POWER-PROCESSING UNIT (PPU)

The task of the PPU in Fig. 4-13 is to supply the desired currents based on the reference signals to the induction motor. This PPU is further illustrated in Fig. 4-14a, where phases b and c are omitted for simplification. One of the easiest ways to ensure that the motor is supplied the desired currents is to use hysteresis control similar to that discussed in Chapter 7 for ECM drives and Chapter 10 for PMAC drives in the previous course [1]. The measured phase current is compared with its

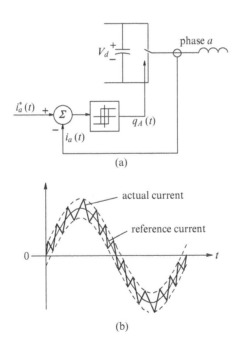

Fig. 4-14 (a) Block diagram representation of hysteresis current control; (b) current waveform.

reference value in the hysteresis comparator, whose output determines the switch state (up or down), resulting in a current waveform as shown in Fig. 4-14b.

In spite of the simplicity of the hysteresis control, one perceived drawback of this controller is that its switching frequency changes as a function of the back-emf waveform. For this reason, constant switching frequency PPU are used as described in Chapter 8 of this book.

4-8   SUMMARY

In this chapter, we have qualitatively examined how it is possible to produce a step in torque in a squirrel-cage induction machine. This analysis is aided by the steady-state analysis of induction machines using space vectors in the previous course, which clearly shows the orthogonal relationship between the rotor flux-linkage space vector and the rotor mmf space vector.

In vector control, we keep the rotor flux linkage constant in amplitude (which can be decreased in the flux-weakening mode). Controlling the rotor flux linkage requires a dq winding analysis, where the current in the equivalent d winding of the stator is kept constant in order to keep the rotor flux along the d-axis constant. A step change in the stator q winding current suddenly induces currents in the rotor equivalent q-axis winding while keeping its flux linkage zero. Therefore, the rotor flux linkage remains unchanged in amplitude. Sudden appearance of currents along the rotor q-axis, in the presence of d-axis flux, results in a step change in torque. To maintain the induced rotor q winding current from decaying, the dq winding set must be rotated at an appropriate slip speed with respect to the rotor.

REFERENCES

1. N. Mohan, *Electric Machines and Drives: A First Course*, Wiley, Hoboken, NJ, 2011. http://www.wiley.com/college/mohan.
2. A. Hughes, J. Corda, and D. Andrade, "Vector Control of Cage Induction Motors: A Physical Insight," IEE Proceedings: Electric Power Applications, vol. 143, no. 1, Jan. 1996, pp. 59–68.

PROBLEMS

4-1 Draw the dynamic dq-axis equivalent circuits under the condition that the rotor flux is aligned with d-axis, such that λ_{rq} and $d\lambda_{rq}/dt$ are zero at all times. Apply a step change of current in the q-axis circuit and explain what happens.

4-2 In an induction motor described with the following nameplate data, establishing the rated air gap flux density requires $\hat{I}_m = 2.54$ A. To build up to this rated flux, calculate the three-phase currents at $t = 0^-$.

Nameplate Data

Power:	3 HP/2.4 kW
Voltage:	460V (L-L, rms)
Frequency:	60 Hz
Phases:	3
Full Load Current:	4A (rms)
Full-Load Speed:	1750 rpm
Full-Load Efficiency:	88.5%
Power Factor:	80.0%
Number of Poles:	4

Per-Phase Motor Circuit Parameters:

$R_s = 1.77\,\Omega$

$R_r = 1.34\,\Omega$

$X_{\ell s} = 5.25\,\Omega$ (at 60 Hz)

$X_{\ell r} = 4.57\,\Omega$ (at 60 Hz)

$X_m = 139.0\,\Omega$ (at 60 Hz)

Full-Load Slip $= 1.72\%$

The iron losses are specified as 78 W and the mechanical (friction and windage) losses are specified as 24 W. The inertia of the machine is given. Assuming that the reflected load inertia is approximately the same as the motor inertia, the total equivalent inertia of the system is $J_{eq} = 0.025\,\text{kg} \cdot \text{m}^2$.

4-3 In the machine of Problem 4-2, a desired step-torque at $t = 0^+$ requires step change in $i_{sq} = 2\,\text{A}$ from its initial zero value. Calculate the phase currents at $t = 0^+$, which result in the desired step change in q-axis current while maintaining the rated flux density in the air gap.

4-4 In the machine of Problems 4-1 and 4-2, the slip speed at which the equivalent d-axis and the q-axis windings need to be rotated is $\omega_{slip} = 2.34$ electrical rad/s. Assuming that the rotor is blocked from turning, calculate the phase currents at $t = 8\,\text{ms}$.

4-5 Repeat Problem 4-4, if the rotor is turning and the speed can be assumed constant at 1100 rpm even after the step change in torque at $t = 0^+$.

5 Mathematical Description of Vector Control in Induction Machines

In vector control described qualitatively in Chapter 4, the *d*-axis is aligned with the rotor flux linkage space vector such that the rotor flux linkage in the *q*-axis is zero. With this as the motivation, we will first develop a model of the induction machine where this condition is always met. Such a model of the machine would be valid regardless if the machine is vector controlled, or if the voltages and currents are applied as under a general-purpose operation (line-fed or in adjustable speed drives described in the previous course).

After developing the motor model, we will study vector control of induction-motor drives, assuming that the exact motor parameters are known—effects of errors in parameter estimates are discussed in the next chapter. We will first use an idealized current-regulated PWM (CR-PWM) inverter to supply motor currents calculated by the controller. As the last step in this chapter, we will use an idealized space vector pulse width-modulated inverter (discussed in detail in Chapter 7) to supply motor voltages that result in the desired currents calculated by the controller.

5-1 MOTOR MODEL WITH THE *d*-AXIS ALIGNED ALONG THE ROTOR FLUX LINKAGE $\vec{\lambda}_r$-AXIS

As noted in the qualitative description of vector control, we will align the *d*-axis (common to both the stator and the rotor) to be along the rotor flux linkage $\vec{\lambda}_r$ ($= \hat{\lambda}_r e^{j0}$), as shown in Fig. 5-1. Therefore,

Advanced Electric Drives: Analysis, Control, and Modeling Using MATLAB/Simulink®, First Edition. Ned Mohan.

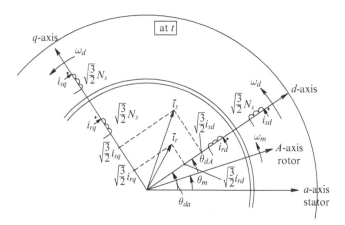

Fig. 5-1 Stator and rotor mmf representation by equivalend dq winding currents. The d-axis is aligned with $\hat{\lambda}_r$.

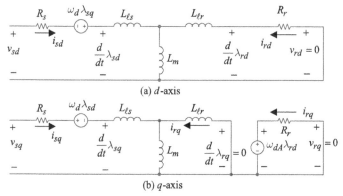

(a) d-axis

(b) q-axis

Fig. 5-2 Dynamic circuits with the d-axis aligned with $\vec{\lambda}_r$.

$$\lambda_{rq}(t) = 0. \tag{5-1}$$

Equating λ_{rq} in Eq. (3-22) to zero,

$$i_{rq} = -\frac{L_m}{L_r} i_{sq}. \tag{5-2}$$

The condition that the d-axis is always aligned with $\vec{\lambda}_r$ such that $\lambda_{rq} = 0$ also results in $d\lambda_{rq}/dt$ to be zero. Using $\lambda_{rq} = 0$ and $d\lambda_{rq}/dt = 0$ in the d- and the q-axis dynamic circuits, we can obtain the simplified circuits shown in Fig. 5-2a and b. Note that Eq. (5-2) is consistent with the equivalent circuit of Fig. 5-2b, where $L_r = L_{\ell r} + L_m$.

Next, we will calculate the slip speed ω_{dA} and the electromagnetic torque T_{em} in this new motor model in terms of the rotor flux λ_{rd} and the stator current component i_{sq} in the *q*-winding (under vector control conditions, λ_{rd} would be kept constant except in the field-weakening mode and the toque production will be controlled by i_{sq}).

We will also establish the dynamics of the rotor flux λ_{rd} in the rotor *d*-winding (λ_{rd} varies during flux buildup at startup and when the motor is made to go into the flux weakening mode of operation).

5-1-1 Calculation of ω_{dA}

As discussed earlier, under the condition that the *d*-axis is always aligned with the rotor flux, the *q*-axis rotor flux linkage is zero, as well as $d\lambda_{rq}/dt = 0$. Therefore, in a squirrel-cage rotor with $v_{rq} = 0$, Eq. (3-32) results in

$$\omega_{dA} = -R_r \frac{i_{rq}}{\lambda_{rd}}, \tag{5-3}$$

which is consistent with the equivalent circuit of Fig. 5-2b. In the rotor circuit, the time-constant τ_r, called the rotor time-constant, is

$$\tau_r = \frac{L_r}{R_r}. \tag{5-4}$$

Substituting for i_{rq} from Eq. (5-2), in terms of τ_r, the slip speed can be expressed as

$$\omega_{dA} = \frac{L_m}{\tau_r \lambda_{rd}} i_{sq}. \tag{5-5}$$

5-1-2 Calculation of T_{em}

Since the flux linkage in the *q*-axis of the rotor is zero, the electromagnetic torque is produced only by the *d*-axis flux in the rotor acting on the rotor *q*-axis winding. Therefore, from Eq. (3-46),

$$T_{em} = -\frac{p}{2} \lambda_{rd} i_{rq}. \tag{5-6}$$

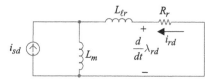

Fig. 5-3   The *d*-axis circuit simplified with a current excitation.

In Eq. (5-6), substituting for i_{rq} from Eq. (5-2)

$$T_{em} = \frac{p}{2}\lambda_{rd}\left(\frac{L_m}{L_r}i_{sq}\right). \tag{5-7}$$

5-1-3   *d*-Axis Rotor Flux Linkage Dynamics

To obtain the dependence of λ_{rd} on i_{sd}, we will make use of the equivalent circuit in Fig. 5-2a, and redraw it as in Fig. 5-3 with a current excitation by i_{sd}. From Fig. 5-3, in terms of Laplace domain variables,

$$i_{rd}(s) = -\frac{sL_m}{R_r + sL_r}i_{sd}(s). \tag{5-8}$$

In the rotor *d*-axis winding, from Eq. (3-21),

$$\lambda_{rd} = L_r i_{rd} + L_m i_{sd}. \tag{5-9}$$

Substituting for i_{rd} from Eq. (5-8) into Eq. (5-9), and using τ_r from Eq. (5-4),

$$\lambda_{rd}(s) = \frac{L_m}{(1+s\tau_r)}i_{sd}(s). \tag{5-10}$$

In time domain, the rotor flux linkage dynamics expressed by Eq. (5-10) is as follows:

$$\frac{d}{dt}\lambda_{rd} + \frac{\lambda_{rd}}{\tau_r} = \frac{L_m}{\tau_r}i_{sd}. \tag{5-11}$$

5-1-4   Motor Model

Based on above equations, a block diagram of an induction-motor model, where the *d*-axis is aligned with the rotor flux linkage, is shown

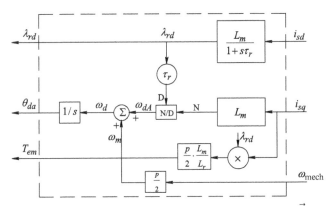

Fig. 5-4 Motor model with d-axis aligned with $\vec{\lambda}_r$.

in Fig. 5-4. The currents i_{sd} and i_{sq} are the inputs, and λ_{rd}, θ_{da}, and T_{em} are the outputs. Note that ω_d $(= \omega_{dA} + \omega_m)$ is the speed of the rotor field, and therefore, the rotor-field angle with respect to the stator *a*-axis (see Fig. 5-1) is

$$\theta_{da}(t) = 0 + \int_0^t \omega_d(\tau)d\tau, \qquad (5\text{-}12)$$

where τ is the variable of integration, and the initial value of θ_{da} is assumed to be zero at $t = 0$.

EXAMPLE 5-1

The motor model developed earlier, with the *d*-axis aligned with $\vec{\lambda}_r$, can be used to model induction machines where vector control is not the objective. To illustrate this, we will repeat the simulation of Example 3-3 of a line-fed motor using this new motor model (which is much simpler) and compare simulation results of these two examples.

Solution

We need to recalculate initial flux values because now the rotor flux linkage is completely along the *d*-axis. This is done in a MATLAB
(Continued)

file EX5_1calc.m on the accompanying website. The initial part in this file is the same as in EX3_1.m (used in Example 3-3), in which the initial values of the angles *thetar* and *thetas* for $\vec{\lambda}_r$ and $\vec{\lambda}_s$ are calculated with respect to the *d*-axis aligned to the stator *a*-axis with $\theta_{da}(0) = 0$. In the present model, with the *d*-axis aligned with $\vec{\lambda}_r$, the rotor flux linkage angle is zero, and the stator flux linkage angle with respect to the *d*-axis equals (*thetas*−*thetar*) in terms of their values in EX3.1m.

The Simulink schematic for this example is called EX5_1.mdl (included on the accompanying website) and its top-level diagram is shown in Fig. 5-5. The resulting torque and speed plots due to a load torque disturbance in this line-fed machine are plotted in Fig. 5-6, which are identical to the results obtained in Example 3-3.

5-2 VECTOR CONTROL

One of the vector control methods is discussed in this section. It is called indirect vector control in the rotor flux reference frame. For many other possible methods and their pros and cons, readers are urged to look at several books on vector control and the IEEE transactions and conference proceedings of its various societies.

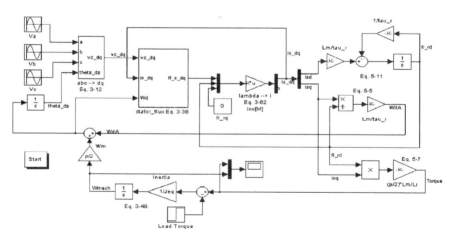

Fig. 5-5   Simulation of Example 5-1.

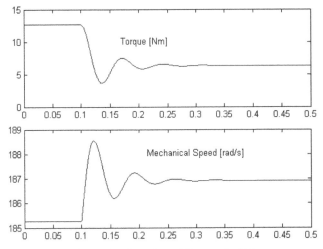

Fig. 5-6   Results of Example 5-1.

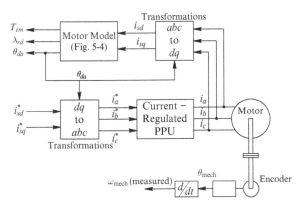

Fig. 5-7   Vector-controlled induction motor with a CR-PWM inverter.

A partial block diagram of a vector-controlled induction motor drive is shown in Fig. 5-7, with the two reference (or command) currents indicated by "*" as inputs. The d-winding reference current i_{sd}^* controls the rotor flux linkage λ_{rd}, whereas the q-winding current i_{sq}^* controls the electromagnetic torque T_{em} developed by the motor. The reference dq winding currents (the outputs of the proportional-integral PI controllers described in the next section) are converted into the reference phase currents $i_a^*(t)$, $i_b^*(t)$, and $i_c^*(t)$. A current-regulated switch-mode converter (the power-processing unit, PPU) can deliver the desired currents to the induction motor, using a tolerance-band control described in the previous chapter. However, in such a current-regulation scheme, the switching frequency within the PPU does not remain constant. If it is important to keep this switching frequency constant, then an alternative is described in a Section 5-3 using a space-vector pulse-width-modulation scheme, which is discussed in detail in Chapter 8.

5-2-1   Speed and Position Control Loops

The current references i_{sd}^* and i_{sq}^* (inputs in the block diagram of Fig. 5-7) are generated by the cascaded torque, speed, and position control loops shown in the block diagram of Fig. 5-8, where θ_{mech}^* is the position reference input. The actual position θ_{mech} and the rotor speed ω_{mech} (where $\omega_m = p/2 \times \omega_{mech}$) are measured, and the rotor flux linkage λ_{rd} is calculated as shown in the block diagram of Fig. 5-8 (same as Fig. 4-13 of the previous chapter). For operation in an extended speed range beyond the rated speed, the flux weakening is implemented as a function of rotor speed in computing the reference for the rotor flux linkage.

EXAMPLE 5-2

In this example, we will consider the drive system of Example 5-1 under vector control described earlier. The initial conditions in the motor are identical to that in the previous example. We will neglect the torque loop in this example, where all the motor parameter estimates are assumed to be perfect. (We will see the effect of estimate errors in the motor parameters in the next chapter.) The objective of the speed loop is to keep the speed at its initial value, in spite of the load torque disturbance at $t = 0.1$ second. We will design the speed loop with a bandwidth of 25 rad/s and a phase margin of 60°, using the same procedure as in Reference [1].

Solution

Initial flux values are the same as in Example 5-1. These calculations are repeated in a MATLAB file EX5_2calc.m on the accompanying website. To design the speed loop (without the torque loop), the torque expression is derived as follows at the rated value of i_{sd}^*: In

---

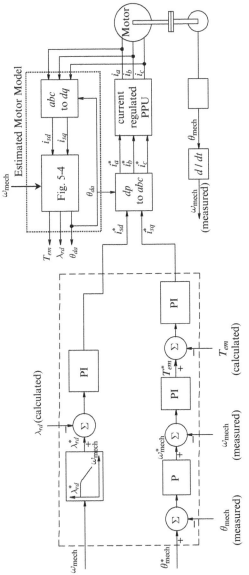

Fig. 5-8 Vector controlled induction motor drive with a current-regulated PPU.

87

steady state under vector control, $i_{rd} = 0$ in Fig. 5-3. Therefore, in Eq. (5-9)

$$\lambda_{rd} = L_m i_{sd} \quad \text{(under vector control in steady state).} \quad (5\text{-}13)$$

Substituting for λ_{rd} from Eq. (5-13) and for i_{rq} from Eq. (5-2) into the torque expression of Eq. (5-7) at the rated i_{sd}^*,

$$T_{em} = \underbrace{\frac{p}{2} \frac{L_m^2}{L_r} i_{sd}^*}_{k} i_{sq} \quad \text{(under vector control in steady state),} \quad (5\text{-}14)$$

where k is a constant. The speed loop diagram is shown in Fig. 5-9 where the PI controller constants are calculated in EX5_2calc.m on the basis that the crossover frequency of the open loop is 25 rad/s and the phase margin is 60°.

The simulation diagram of the file EX5_2.mdl (included on the accompanying website) is shown in Fig. 5-10, and the torque and speed are plotted in Fig. 5-11.

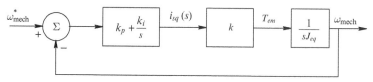

Fig. 5-9   Design of the speed-loop controller.

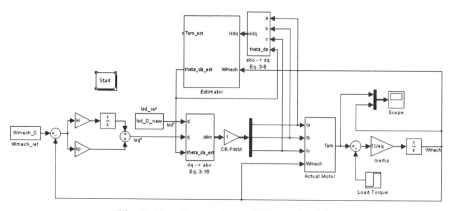

Fig. 5-10   Simulation of Example 5-2.

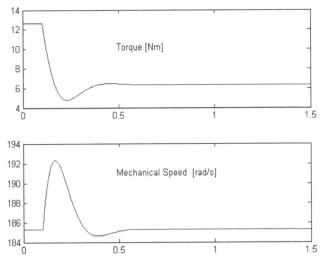

Fig. 5-11   Simulation results of Example 5-2.

5-2-2   Initial Startup

Unlike the example above where the system was operating in steady state initially, the system must be started from standstill conditions. Initially, the flux is built up to its rated value, keeping the torque to be zero. Therefore, initially i_{sq}^* is zero. The reference value λ_{rd}^* of the rotor flux at zero speed is calculated in the block diagram of Fig. 5-8. The value of the rotor-field angle θ_{da} is assumed to be zero. The division by zero in the block diagram of Fig. 5-4 is prevented until λ_{rd} takes on some finite (nonzero) value. This way, three stator currents build up to their steady state dc magnetizing values. The rotor flux builds up entirely along the a-axis. Once the dynamics of the flux build-up is completed, the drive is ready to follow the torque, speed and position commands.

5-2-3   Calculating the Stator Voltages to Be Applied

It is usually desirable to keep the switching frequency within the switch-mode converter (PPU) constant. Therefore, it is a common practice to calculate the required stator voltages that the PPU must supply to the motor, in order to make the stator currents equal to their reference values.

We will first define a unitless term called the leakage factor σ of the induction machine as:

$$\sigma = 1 - \frac{L_m^2}{L_s L_r}. \tag{5-15}$$

Substituting for i_{rd} from Eq. (5-9) into Eq. (3-19) for λ_{sd},

$$\lambda_{sd} = \sigma L_s i_{sd} + \frac{L_m}{L_r} \lambda_{rd}. \tag{5-16}$$

From Eq. (3-20) for λ_{sq}, using Eq. (5-2) under vector-controlled conditions

$$\lambda_{sq} = \sigma L_s i_{sq}. \tag{5-17}$$

Substituting these into Eq. (3-28) and Eq. (3-29) for v_{sd} and v_{sq},

$$v_{sd} = \underbrace{R_s i_{sd} + \sigma L_s \frac{d}{dt} i_{sd}}_{v'_{sd}} + \underbrace{\frac{L_m}{L_r} \frac{d}{dt} \lambda_{rd} - \omega_d \sigma L_s i_{sq}}_{v_{sd,comp}} \tag{5-18}$$

and

$$v_{sq} = \underbrace{R_s i_{sq} + \sigma L_s \frac{d}{dt} i_{sq}}_{v'_{sq}} + \underbrace{\omega_d \frac{L_m}{L_r} \lambda_{rd} + \omega_d \sigma L_s i_{sd}}_{v_{sq,comp}}. \tag{5-19}$$

5-2-4   Designing the PI Controllers

In the d-axis voltage equation of Eq. (5-18), on the right side only the first two terms are due to the d-axis current i_{sd} and di_{sd}/dt. The other terms due to λ_{rd} and i_{sq} can be considered as disturbances. Similarly in Eq. (5-19), the terms due to λ_{rd} and i_{sd} can be considered as disturbances. Therefore, we can rewrite these equations as:

$$v'_{sd} = R_s i_{sd} + \sigma L_s \frac{d}{dt} i_{sd} \tag{5-20}$$

and

$$v'_{sq} = R_s i_{sq} + \sigma L_s \frac{d}{dt} i_{sq}, \tag{5-21}$$

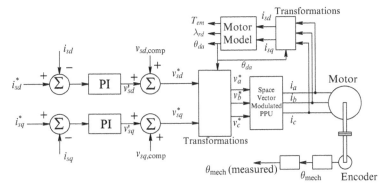

Fig. 5-12   Vector control with applied voltages.

where the compensation terms are

$$v_{sd,comp} = \frac{L_m}{L_r}\frac{d}{dt}\lambda_{rd} - \omega_d\sigma L_s i_{sq} \tag{5-22}$$

and

$$v_{sq,comp} = \omega_d\left(\frac{L_m}{L_r}\lambda_{rd} + \sigma L_s i_{sd}\right). \tag{5-23}$$

As shown in the block diagram of Fig. 5-12, we can generate the reference voltages v_{sd}^* and v_{sq}^* from given i_{sd}^* and i_{sq}^*, and using the calculated values of λ_{rd}, i_{sd}, and i_{sq}, and the chosen value of ω_d. Using the calculated value of θ_{da} in the block diagram of Fig. 5-12, the reference values v_a^*, v_b^*, and v_c^* for the phase voltages are calculated. The actual stator voltages v_a, v_b, and v_c are supplied by the power electronics converter, using the stator voltage space vector modulation technique discussed in Chapter 8.

To obtain v_{sd}' and v_{sq}' signals in Fig. 5-12, we will employ PI controllers in the current loops. To compute the gains of the proportional and the integral portions of the PI controllers, we will assume that the compensation is perfect. Hence, each channel results in a block diagram of Fig. 5-13 (shown for d-axis), where the "motor-load plant" can be represented by the transfer functions below, based on Eq. (5-20) and Eq. (5-21):

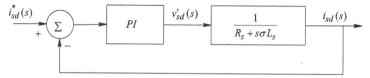

Fig. 5-13 Design of the current-loop controller.

$$i_{sd}(s) = \frac{1}{R_s + s\sigma L_s} v'_{sd}(s) \tag{5-24}$$

and

$$i_{sq}(s) = \frac{1}{R_s + s\sigma L_s} v'_{sq}(s). \tag{5-25}$$

Now, the gain constants of the PI controller in Fig. 5-13 (same in the q-winding) can be calculated using the procedure illustrated in the following example.

EXAMPLE 5-3

Repeat the vector control of Example 5-2 by replacing the CR-PWM inverter by a space vector pulse width-modulated inverter, which is assumed to be ideal. The speed loop specifications are the same as in Example 5-2. The current (torque) loop to generate reference voltage has 10 times the bandwidth of the speed loop and the same phase margin of 60°.

Solution

Calculations for the initial conditions are repeated in the MATLAB file EX5_3calc.m, which is included on the accompanying website. It also shows the procedure for calculating the gain constants of the PI controller of the current loop. The simulation diagram of the SIMU-LINK file EX5_3.mdl (included on the accompanying website) is shown in Fig. 5-14, and the simulation results are plotted in Fig. 5-15.

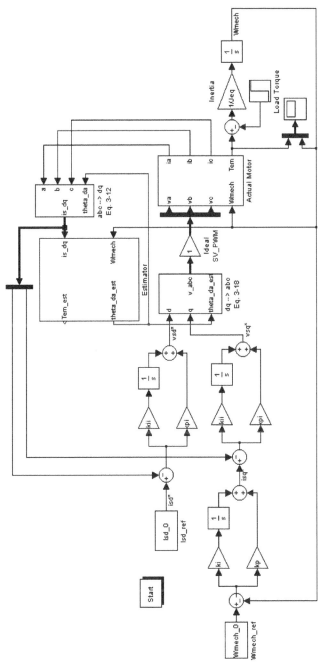

Fig. 5-14 Simulation of Example 5-3.

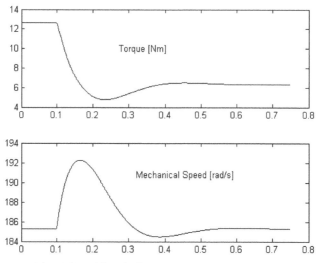

Fig. 5-15   Simulation results of Example 5-3.

EXAMPLE 5-4

Consider the "test" machine described in Chapter 1. This machine is operating in steady state under its rated conditions, supplying its rated torque. At $t = 1$ second, the load torque suddenly goes to one-half of its initial value and stays there. The objective is to maintain the speed of this machine at its initial value.

Design a vector control scheme with the d-axis aligned to the rotor flux. Design the speed loop to have a bandwidth of 25 rad/s and the phase margin of 60°. Design the torque (current) loop to have a bandwidth of 250 rad/second and a phase margin of 60°.

Simulate the system with and without the compensation terms and plot various quantities as functions of time.

Solution

See the complete solution on the accompanying website.

5-3 SUMMARY

In this chapter, we first developed a model of the induction machine where the d-axis is always aligned with the rotor flux linkage space vector. Such a model of the machine is valid regardless if the machine is vector controlled, or if the voltages and currents to it are applied as under a general-purpose operation as discussed in Chapter 3. This is illustrated by Example 5-1.

After developing the earlier-mentioned motor model, we studied vector control of induction motor drives, assuming that the exact motor parameters are known—effects of errors in parameter estimates are discussed in the next chapter. We first used an idealized CR-PWM inverter to supply motor currents calculated by the controller. This vector control is illustrated by means of Example 5-2.

As the last step in this chapter, we used an idealized space vector pulse-width-modulated inverter (discussed in detail in Chapter 8) to supply motor voltages that result in the desired currents calculated by the controller. This is illustrated by means of Example 5-3.

REFERENCE

1. N. Mohan, *Electric Machines and Drives: A First Course*, Wiley, Hoboken, NJ, 2011. http://www.wiley.com/college/mohan.

PROBLEMS

5-1 In Example 5-1, comment how i_{sd}, i_{sq}, and $\hat{\lambda}_r$ vary under the dynamic condition caused by the change in load torque. Plot and comment on ω_d under steady state, as well as under dynamic conditions.

5-2 Modify the simulation of Example 5-1 for a line start from stand-still at $t = 0$, with the rated load torque.

5-3 In Example 5-2, plot the stator dq winding currents, the phase currents, ω_d, and ω_{dA}.

5-4 Add the blocks necessary in the simulation of Example 5-2 to plot phase voltages.

5-5 In the simulation of Example 5-2, include the torque loop, assuming its bandwidth to be 10 times larger than the speed loop bandwidth of 25 rad/s (keeping the phase margin in both loops at 60°). Compare results with those in Example 5-2.

5-6 Include flux weakening in the simulation of Example 5-2 by modifying the simulation as follows: initially in the steady-state operating condition, the load torque is one-half the rated torque of the motor. Instead of the load disturbance at $t = 0.1$ second, the speed reference is ramped linearly to reach 1.5 times the full-load motor speed in 2 seconds.

5-7 Plot phase voltages in Example 5-3.

5-8 Add the compensation terms in the simulation of Example 5-3. Compare results with those of Example 5-3.

5-9 Repeat Problem 5-6 in the simulation of Example 5-3 by including flux weakening, 0.1 second, the speed reference.

6 Detuning Effects in Induction Motor Vector Control

In vector control described in Chapters 4 and 5, we assumed that the induction machine parameters were known exactly. In practice, the estimated parameters may be off by a significant amount. This is particularly true of the rotor time constant τ_r $(=L_r/R_r)$, which depends on the rotor resistance that increases significantly as the rotor heats up. In this chapter, we will calculate the steady-state error due to the incorrect estimate of the rotor resistance and also look at its effect on the dynamic response of vector-controlled drives [1,2].

6-1 EFFECT OF DETUNING DUE TO INCORRECT ROTOR TIME CONSTANT τ_r

We will define a detuning factor to be the ratio of the actual and the estimated rotor time constants as

$$k_\tau = \frac{\tau_r}{\tau_{r,\text{est}}}, \tag{6-1}$$

where the estimated quantities are indicated by the subscript "est." To analytically study the sensitivity of the vector control to k_τ, we will simplify our system by assuming that the rotor of the induction machine is blocked from turning, that is, $\omega_{\text{mech}} = 0$. Also, we will assume an open-loop system, where the command (reference) currents are i_{sd}^* and i_{sq}^*.

As shown in Fig. 6-1 at $t = 0^-$, the stator a-axis, the rotor A-axis, and the d-axis are all aligned with $\vec{\lambda}_r$, which is built up to its rated value.

Advanced Electric Drives: Analysis, Control, and Modeling Using MATLAB/Simulink®, First Edition. Ned Mohan.
© 2014 John Wiley & Sons, Inc. Published 2014 by John Wiley & Sons, Inc.

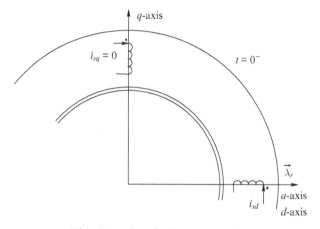

Fig. 6-1   *dq* windings at $t = 0^-$.

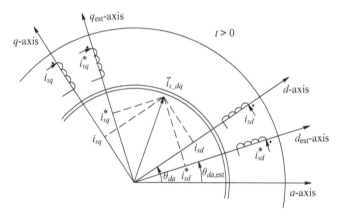

Fig. 6-2   *dq* windings at $t > 0$; drawn for $k_\tau < 1$.

Also initially, $i_{sd} = i_{sd}^*$, and $\theta_{da} = \theta_{da,\text{est}} = 0$. Initially, the torque component of the stator current is assumed to be zero, that is $i_{sq}^* = i_{sq} = 0$. At
$t = 0^+$ and beyond, i_{sd}^* remains unchanged, but there is a step jump in
i_{sq}^* to produce a step change in torque.

The commanded currents are supplied to the estimated *dq* windings
shown in Fig. 6-2, drawn for $k_\tau < 1$ at some time *t*. On the basis of dis-
cussions later on, we will show that for $k_\tau < 1$, $\theta_{da,\text{est}}$ will be less than
the actual θ_{da}. The angle θ_{da} is the angle of the actual *d*-axis along which
the rotor flux-linkage vector lies (*not* the estimated axis, which the

controller considers, incorrectly of course, to be the actual d-axis). Therefore, with the help of Fig. 6-2, we can compute the currents in the dq stator windings (along the correct dq axes in the actual motor) by projecting the dq winding currents along the estimated axes

$$\vec{i}_{s\_dq}^{d} = \vec{i}_{s\_dq}^{d,\text{est}} e^{-j(\theta_{da} - \theta_{da,\text{est}})} = \vec{i}_{s\_dq}^{d,\text{est}} e^{-j(\theta_{\text{err}})}, \tag{6-2}$$

where

$$\theta_{\text{err}} = \theta_{da} - \theta_{da,\text{est}}. \tag{6-3}$$

Note that in this detuned case, we are applying (although mistakenly) the commanded currents to the windings along the estimated d–q axes. Therefore,

$$\vec{i}_{s\_dq}^{d,\text{est}} = i_{sd}^{*} + j i_{sq}^{*}. \tag{6-4}$$

Using Eq. (6-4) in Eq. (6-2),

$$\vec{i}_{s\_dq}^{d} = (i_{sd}^{*} + j i_{sq}^{*}) e^{-j(\theta_{\text{err}})}, \tag{6-5}$$

where $\vec{i}_{s\_dq}^{d} = i_{sd} + j i_{sq}$. Therefore, in Eq. (6-5), the currents in the d- and q-axis windings (in the actual motor) can be calculated as

$$i_{sd} = i_{sd}^{*} \cos\theta_{\text{err}} + i_{sq}^{*} \sin\theta_{\text{err}} \tag{6-6}$$

and

$$i_{sq} = i_{sq}^{*} \cos\theta_{\text{err}} - i_{sd}^{*} \sin\theta_{\text{err}}. \tag{6-7}$$

Figure 6-3 shows the block diagram under the blocked-rotor condition, where the feedback controller is omitted for clarity; instead, i_{sd}^{*} and i_{sq}^{*} are the command (reference) inputs. The torque command is applied at time $t = 0$, assuming that the rotor flux prior to that has built up to the correct value λ_{rd} ($= L_{m} i_{sd}^{*}$) and both θ_{da} and $\theta_{da,\text{est}}$ are initially equal to zero. Assuming that the CR-PWM inverter in Fig. 6-3 is ideal, the currents in the windings along the estimated axis-set equal the commanded values, that is $i_{sd,\text{est}} = i_{sd}^{*}$ and $i_{sq,\text{est}} = i_{sq}^{*}$ in the estimated motor model. The estimated slip speed $\omega_{dA,\text{est}}$ ($= \omega_{d,\text{est}}$ since ω_{mech} equals 0) immediately reaches its steady state value, which is an incorrect value due to the error in the rotor time constant τ_{r}.

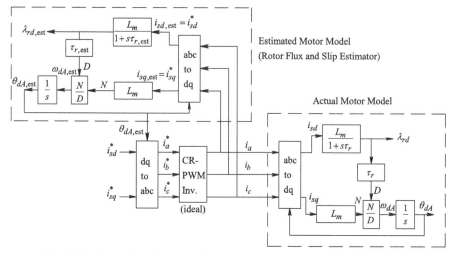

Fig. 6-3   Actual and the estimated motor models (blocked-rotor).

In the actual motor model, modeled in the rotor-flux frame (with the actual d-axis aligned with $\vec{\lambda}_r$) with the correct τ_r, the currents in the dq-windings go through dynamics, as shown in the following example.

EXAMPLE 6-1

In an induction motor, with the nameplate data given below, under a blocked-rotor condition, the flux is initially built up to its rated value (with $i_{sd}(0) = 3.1\,\text{A}$). The toque is commanded to change as a step ($i_{sq}^* = 4.0\,\text{A}$) from zero to nearly 50% of its rated value. Plot the dynamics of i_{sd}, i_{sq}, slip speed ω_{dA}, and the error θ_{err} as functions of time if all the motor parameters are estimated correctly except the rotor resistance is estimated to be one-half its actual value.

Nameplate Data

Power:	3 HP/2.4 kW
Voltage:	460 V (L-L, rms)
Frequency:	60 Hz
Phases:	3
Full-Load Current:	4 A

Full-Load Speed: 1750 rpm

Full-Load Efficiency: 88.5%

Power Factor: 80.0%

Number of Poles: 4

Per-Phase Motor Circuit Parameters

$R_s = 1.77\,\Omega$

$R_r = 1.34\,\Omega$

$X_{\ell s} = 5.25\,\Omega\,(\text{at } 60\,\text{Hz})$

$X_{\ell r} = 4.57\,\Omega\,(\text{at } 60\,\text{Hz})$

$X_m = 139.0\,\Omega\,(\text{at } 60\,\text{Hz})$

Full-Load Slip $= 1.72\%$

The iron losses are specified as 78 W and the mechanical (friction and windage) losses are specified as 24 W. The inertia of the machine is given. Assuming that the reflected load inertia is approximately the same as the motor inertia, the total equivalent inertia of the system is $J_{eq} = 0.025\,\text{kg} \cdot \text{m}^2$.

Solution

Figure 6-4 shows the simulation diagram in Simulink and the dynamics of i_{sd}, i_{sq}, slip speed ω_{dA}, and the error θ_{err} are plotted in Fig. 6-5. We should note that after a dynamic response, each variable comes to a constant steady state value. Also, for $R_{r\_est} = 0.5R_r$, which results in $k_\tau = 0.5$, θ_{err} has a positive value in steady state.

6-2   STEADY-STATE ANALYSIS

We can obtain the results of detuning in steady state by making use of two conditions:

1. *The actual slip speed equals its estimated value in steady state.* We can show this to be the case as follows: In Fig. 6-3, the three-phase currents reach steady state at a frequency of $\omega_{dA,\text{est}}$, and the rotor

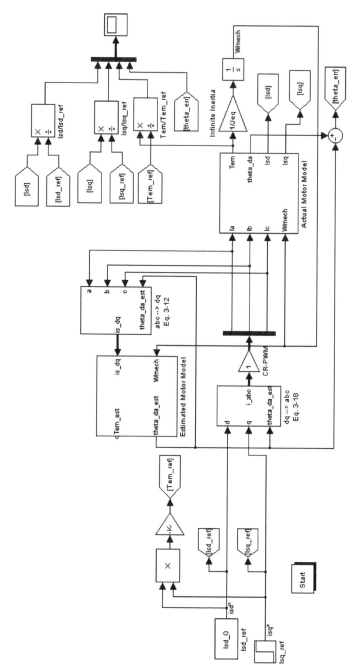

Fig. 6-4 Simulation of Example 6-1.

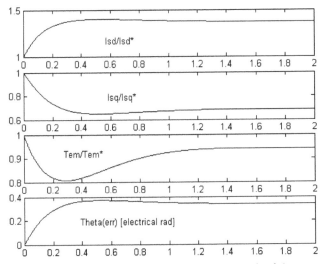

Fig. 6-5   Simulation results of Example 6-1.

flux $\vec{\lambda}_r$ in the actual motor has a constant amplitude and rotates at $\omega_{dA,est}$ in steady state. Therefore, in the actual motor model in the rotor-flux reference frame, modeled with the actual motor parameters, i_{sd} and i_{sq} are dc quantities in steady state. This is possible only if $\omega_{dA} = \omega_{dA,est}$ in steady state.

2. *The magnitude of the stator* dq *current vector is the same in the estimator block and in the actual motor model* — however, the projections are different because the actual and the estimated dq-axes sets are not aligned.

From Eq. (5-5) of the previous chapter repeated here,

$$\omega_{dA} = \frac{1}{\tau_r \lambda_{rd}}(L_m i_{sq}). \tag{6-8}$$

In steady state, the rotor d-winding current is zero ($i_{rd} = 0$ in steady state). Therefore,

$$\lambda_{rd} = L_m i_{sd} \quad \text{(steady state)}. \tag{6-9}$$

Substituting the expression for the rotor flux in steady state from Eq. (6-9) into the slip speed expression of Eq. (6-8),

$$\omega_{dA} = \frac{1}{\tau_r} \frac{i_{sq}}{i_{sd}} \quad \text{(steady state).} \tag{6-10}$$

In the rotor flux and the slip estimator block of Fig. 6-3, the estimated slip speed is based on the commanded currents, which are applied to the windings along the estimated dq-axis:

$$\omega_{dA,est} = \frac{1}{\tau_{r,est}} \frac{i_{sq}^*}{i_{sd}^*} \quad \text{(steady state).} \tag{6-11}$$

Therefore, using condition 1 discussed earlier,

$$\underbrace{\frac{1}{\tau_{r,est}} \frac{i_{sq}^*}{i_{sd}^*}}_{\omega_{dA,est}} = \underbrace{\frac{1}{\tau_r} \frac{i_{sq}}{i_{sd}}}_{\omega_{dA}} \quad \text{(in steady state).} \tag{6-12}$$

Using condition 2 discussed before,

$$\sqrt{i_{sd}^2 + i_{sq}^2} = \sqrt{i_{sd}^{*2} + i_{sq}^{*2}} = \hat{I}_{s\_dq}. \tag{6-13}$$

We will introduce a factor m, called the torque factor, which is the ratio of the commanded toque component to the flux component of the stator current

$$m = \frac{i_{sq}^*}{i_{sd}^*}. \tag{6-14}$$

6-2-1   Steady-State i_{sd}/i_{sd}^*

From Eq. (6-12) and Eq. (6-13) with k_τ given by Eq. (6-1) and m defined in Eq. (6-14),

$$\frac{i_{sd}}{i_{sd}^*} = \sqrt{\frac{1+m^2}{1+k_\tau^2 \cdot m^2}}. \tag{6-15}$$

6-2-2   Steady-State i_{sq}/i_{sq}^*

Similar to the d-axis, in the q-axis,

$$\frac{i_{sq}}{i_{sq}^*} = k_\tau \sqrt{\frac{1+m^2}{1+k_\tau^2 \cdot m^2}}. \tag{6-16}$$

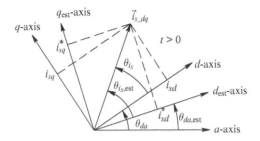

Fig. 6-6 dq winding currents at $t > 0$; drawn for $k_\tau < 1$.

From Eq. (6-15) and Eq. (6-16) in steady state

$$\frac{i_{sq}}{i_{sq}^*} = k_\tau \frac{i_{sd}}{i_{sd}^*}. \tag{6-17}$$

6-2-3   Steady-State θ_{err}

We can calculate the steady state angle error θ_{err} (defined by Eq. 6-3) in Fig. 6-6, noting that

$$\theta_{i_s,est} + \theta_{da,est} = \theta_{i_s} + \theta_{da}. \tag{6-18}$$

Therefore, from Eq. (6-3), using Eq. (6-18)

$$\theta_{err} = \theta_{i_s,est} - \theta_{i_s}, \tag{6-19}$$

where

$$\theta_{i_s,est} = \tan^{-1}\left(\frac{i_{sq}^*}{i_{sd}^*}\right) \tag{6-20}$$

and

$$\theta_{i_s} = \tan^{-1}\left(\frac{i_{sq}}{i_{sd}}\right). \tag{6-21}$$

Hence, in terms of the torque factor defined in Eq. (6-14), and using Eq. (6-17),

$$\theta_{err} = \tan^{-1}(m) - \tan^{-1}(k_\tau \cdot m) \quad \text{(steady state)}. \tag{6-22}$$

6-2-4 Steady-State T_{em}/T_{em}^*

In Eq. (5-7) of Chapter 5, substituting from Eq. (6-9),

$$T_{em} = \left(\frac{p}{2}\frac{L_m^2}{L_r}\right)i_{sd}i_{sq} \quad \text{(steady state)}. \tag{6-23}$$

Therefore, assuming that the estimates of L_m and L_r are correct,

$$\frac{T_{em}}{T_{em}^*} = k_\tau \frac{1+m^2}{1+(k_\tau \cdot m)^2} \quad \text{(steady state)}. \tag{6-24}$$

EXAMPLE 6-2

In the induction motor under the operating conditions described in Example 6-1, all the motor parameters are estimated correctly except the rotor resistance is estimated to be one-half its actual value. Obtain the steady state values of

$$\frac{i_{sd}}{i_{sd}^*}, \quad \frac{i_{sq}}{i_{sq}^*}, \quad \frac{T_{em}}{T_{em}^*}, \quad \text{and} \quad \theta_{err}.$$

Solution

The results obtained using the MATLAB file EX6-2calc.m included on the accompanying website are as follows:

$$\frac{i_{sd}}{i_{sd}^*} = 1.37, \quad \frac{i_{sq}}{i_{sq}^*} = 0.69, \quad \frac{T_{em}}{T_{em}^*} = 0.94, \quad \text{and} \quad \theta_{err} = 0.338 \text{ rad}.$$

All of these values match the steady-state values obtained in Example 6-1.

EXAMPLE 6-3

Consider the "test" machine described in Chapter 1 and considered in Example 5-4 of the previous chapter. Now consider that the estimated rotor resistance is incorrect and $R_{r\_est} = 1.1R_r$.

Modify your simulation in Example 5-4 and simulate the system without the compensation terms. Assume that the load toque is constant at its rated value. Keep the same initial conditions as in Example 5-4. Plot the error in the estimated θ_{da} (error $= \theta_{da} - \theta_{da,est}$) as a function of time. Compare in steady state the amplitudes of the phase currents to the case with no error in the rotor resistance estimate, as was the case in Example 5-4.

Solution

See the complete solution on the accompanying website.

6-3 SUMMARY

In this chapter, we have calculated the steady-state error due to the incorrect estimate of the rotor resistance and also looked at its effect on the dynamic response of vector-controlled drives. The discussion in this chapter is carried out using a blocked-rotor condition, ignoring the feedback loops to show the angle error resulting due to the error in estimation. This discussion can be extended to finite speeds and to the error in the rotor inductance estimation. Another extension of this analysis is to replace the mechanical encoder by a speed estimation block, described in Chapter 8, dealing with the direct torque control (DTC), and to look at the sensitivity of performance to errors in the speed estimation.

REFERENCES

1. D. Novotny and T. Lipo, *Vector Control and Dynamics of AC Drives*, Oxford University Press, New York, 1996.

2. N. Mohan, M. Riaz, and A. Jain, "Explanation of De-Tuning in Vector-Controlled Induction Motor Drives Simplified by Physical-Based Analysis and PSpice Modeling," Proceeding of the European Power Electronics Conference, Lausanne, Switzerland, 1999.

PROBLEMS

6-1 Obtain the results in Example 6-1 for three values of k_r: 0.75, 1.0, and 1.5.

6-2 Repeat the calculations in Example 6-2 and compare results with Problem 6-1 in steady state for k_r: 0.75, 1.0, and 1.5.

7 Dynamic Analysis of Doubly Fed Induction Generators and Their Vector Control

Doubly fed induction generators (DFIGs) are used in harnessing wind energy. The principle of operation of doubly wound induction machines was described in steady state in Reference [1]. In this chapter, we will mathematically describe doubly wound induction machines in order to apply vector control. As an introduction, Fig. 7-1 shows a doubly fed induction generator.

The cross-section of a DFIG is shown in Fig. 7-2. It consists of a stator, similar to the squirrel-cage induction machines, with a three-phase winding, each having N_s turns per phase that are assumed to be distributed sinusoidally in space. The rotor consists of a wye-connected three-phase windings, each having N_r turns per phase that are assumed to be distributed sinusoidally in space. Its terminals, A, B, and C, are supplied appropriate currents through slip-rings and brushes, as shown in Fig. 7-1b.

The benefits of using a DFIG in wind applications are as follows:

1. The speed can be control over a sufficiently wide range to make the turbine operate at its optimum coefficient of performance C_p.

2. The stator is directly connected to the grid. Only the rotor is supplied through power electronics that is approximately rated at 30% of the rated power of the wind turbine.

3. The reactive power supplied to the rotor is controllable and it is amplified on the stator-side.

Advanced Electric Drives: Analysis, Control, and Modeling Using MATLAB/Simulink®, First Edition. Ned Mohan.
© 2014 John Wiley & Sons, Inc. Published 2014 by John Wiley & Sons, Inc.

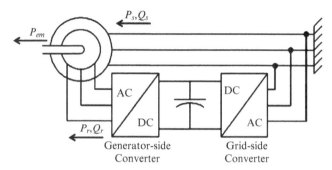

Fig. 7-1 Doubly fed induction generator (DFIG) where P and Q inputs are defined into the stator and the rotor.

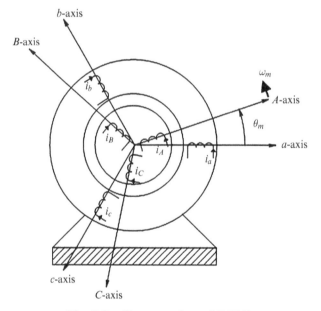

Fig. 7-2 Cross-section of DFIG.

A major disadvantage of DFIGs is the periodic maintenance required of slip-rings and brushes.

7-1 UNDERSTANDING DFIG OPERATION

Prior to writing dynamic equations, we will describe the DFIG operation by first assuming steady state and neglecting all parasitic, such as

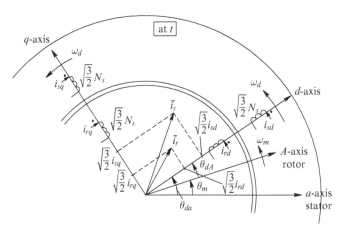

Fig. 7-3 d-axis aligned with the rotor flux; stator and rotor current vectors are shown for definition purposes only.

stator and rotor leakage inductances and resistances. We will assume the number of turns on the stator and the rotor windings to be the same. This operation is described in terms of dq-axis, as compared with the steady-state analysis in Reference [1], which was described without the help of dq-axis analysis. In this analysis, we will assume that the d-axis is aligned with the rotor flux-linkage space vector such that the rotor flux linkage in the q-axis is zero. This is shown in Fig. 7-3.

It should be noted that having neglected the leakage fluxes, the flux-linkage in the rotor d-axis is the same as the stator flux in the d-axis (refer to Eq. 3-19, Eq. 3-20, Eq. 3-22). We will write all the necessary equations in steady state under the assumptions indicated earlier and where P and Q inputs are defined into the stator and the rotor. Using the equations in Chapter, the following equations can be written.

Stator Voltages

$$v_{sd} = -\omega_d \lambda_{sq} = 0 \quad (\text{since } \lambda_{sq} = 0) \tag{7-1}$$

$$v_{sq} = \omega_d \lambda_{sd} \tag{7-2}$$

$$v_{sq} = \omega_d \lambda_{sd} = \sqrt{\frac{2}{3}} \hat{V}_s \equiv \text{constant (since } v_{sd} = 0), \tag{7-3}$$

where $\hat{V}_s$ is the amplitude of the stator voltage space vector.

Flux Linkages and Currents

d-*axis*

$$\lambda_{rd} = \lambda_{sd} = \frac{v_{sq}}{\omega_d} \equiv \text{constant} \tag{7-4}$$

$$\lambda_{sd} = \lambda_{rd} = L_m(i_{sd} + i_{rd}) \tag{7-5}$$

$$i_{sd} = \frac{\lambda_{sd}}{L_m} - i_{rd} = i_{md} - i_{rd}, \tag{7-6}$$

where the magnetizing current $i_{md} = (\lambda_{sd}/L_m)$.

q-*axis*

$$\lambda_{rq} = \lambda_{sq} = 0 \tag{7-7}$$

$$\text{since } \lambda_{sq} = \lambda_{rq} = L_m(i_{sq} + i_{rq}) = 0 \tag{7-8}$$

$$i_{sq} = -i_{rq} \tag{7-9}$$

$$\vec{i}_{s\_dq} = i_{sd} + ji_{sq} = (i_{md} - i_{rd}) - ji_{rq}. \tag{7-10}$$

Rotor Voltages

$$v_{rd} = -\omega_{dA}\lambda_{rq} = 0 \quad (\text{since } \lambda_{rq} = 0) \tag{7-11}$$

$$v_{rq} = \omega_{dA}\lambda_{rd} \tag{7-12}$$

$$v_{rq} = s\omega_d\lambda_{sd}, \tag{7-13}$$

where s is the slip.

Stator and Rotor Power Inputs

Stator

$$P_s + jQ_s = \left(\underbrace{v_{sd}}_{(=0)} + jv_{sq}\right)(i_{sd} - ji_{sq}) = v_{sq}i_{sq} + jv_{sq}i_{sd} \tag{7-14}$$

$$P_s = v_{sq}i_{sq} = \omega_d\lambda_{sd}i_{sq} \tag{7-15}$$

$$Q_s = v_{sq}i_{sd} = \omega_d\lambda_{sd}i_{sd}. \tag{7-16}$$

Rotor

$$P_r + jQ_r = \left(\underbrace{v_{rd}}_{(=0)} + jv_{rq}\right)(i_{rd} - ji_{rq}) = v_{rq}i_{rq} + jv_{rq}i_{rd} \tag{7-17}$$

$$P_r = v_{rq}i_{rq} = s\omega_d\lambda_{rd}i_{rq} \tag{7-18}$$

$$Q_r = v_{rq}i_{rd} = s\omega_d\lambda_{rd}i_{rd}. \tag{7-19}$$

Electromagnetic Torque

$$T_{em} = -\frac{p}{2}\lambda_{rd}i_{rq} \quad \text{(using Eq. 3-46 and } \lambda_{rq} = 0). \tag{7-20}$$

Relationships of Stator and Rotor Real and Reactive Powers

$$\frac{P_s}{P_r} = -\frac{1}{s} \tag{7-21}$$

$$Q_s = (\omega_d L_m i_{md}^2) - \frac{Q_r}{s} = Q_{mag} - \frac{Q_r}{s}. \tag{7-22}$$

EXAMPLE 7-1

A DFIG is operating in the motoring mode at a subsynchronous speed at a lagging power factor (drawing Q_s from the grid). Calculate the signs of various quantities in this mode of operation.

Solution

$$\omega_{slip} = \omega_{dA} = + \tag{7-23}$$

$$s = \frac{\omega_{slip}}{\omega_{syn}} = + \tag{7-24}$$

$$T_{em} = + \tag{7-25}$$

$$P_s = v_{sq}i_{sq} = \omega_d\lambda_{sd}i_{sq} = + \tag{7-26}$$

$$\therefore i_{sq} = + \tag{7-27}$$

(*Continued*)

$$i_{rq} = -$$ (7-28)

$$Q_s = v_{sq}i_{sd} = \omega_d\lambda_{sd}i_{sd} = \underbrace{+}_{\text{given}}$$ (7-29)

$$\therefore i_{sd} = +$$ (7-30)

$$P_r = v_{rq}i_{rq} = s\omega_d\lambda_{rd}i_{rq} = -$$ (7-31)

$$Q_r = v_{rq}i_{rd} = s\omega_d\lambda_{rd}i_{rd}$$

$$i_{rq} = -\ (\text{since } i_{sq} = +)$$ (7-32)

$$i_{rd} = +\ \text{taken as positive but } i_{rd} < i_{md}$$ (7-33)

$$v_{rd} = 0$$ (7-34)

$$v_{rq} = s\omega_d\lambda_{rd} = +$$ (7-35)

$$\begin{bmatrix} v_A(t) \\ v_B(t) \\ v_C(t) \end{bmatrix} = \sqrt{\frac{2}{3}} \begin{bmatrix} \cos(\theta_{dA}) & -\sin(\theta_{dA}) \\ \cos\left(\theta_{dA} + \dfrac{4\pi}{3}\right) & -\sin\left(\theta_{dA} + \dfrac{4\pi}{3}\right) \\ \cos\left(\theta_{dA} + \dfrac{2\pi}{3}\right) & -\sin\left(\theta_{dA} + \dfrac{2\pi}{3}\right) \end{bmatrix} \begin{bmatrix} v_{rd} \\ v_{rq} \end{bmatrix}.$$ (7-36)

Various space vectors are shown in Fig. 7-4.

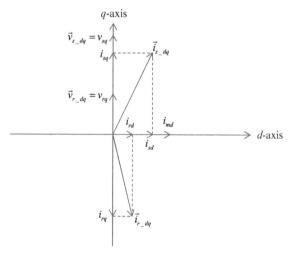

Fig. 7-4   Space vector diagram for Example 7-1.

EXAMPLE 7-2

A DFIG is operating in the generator mode at a supersynchronous speed at a leading power factor (supplying Q_s to the grid). Calculate the signs of various quantities in this mode of operation (Fig. 7-5).

Solution

$$\omega_{slip} = \omega_{dA} = -$$

$$s = \frac{\omega_{slip}}{\omega_{syn}} = -$$

$$T_{em} = -$$

$$P_s = -$$

$$P_s = v_{sq}i_{sq} = \omega_d\lambda_{sd}i_{sq}$$

$$i_{sq} = -$$

$$Q_s = v_{sq}i_{sd} = \omega_d\lambda_{sd}i_{sd} = \underset{given}{-}$$

$$i_{sd} = -$$

$$Q_s = Q_{mag} - \frac{Q_r}{s} = - \therefore Q_r = - \text{ such that } \frac{Q_r}{s} > Q_{mag}$$

$$\omega_{dA} = \omega_{syn} - \omega_m = -$$

$$i_{rq} = + (\text{since } i_{sq} = -)$$

$$P_r = v_{rq}i_{rq} = \omega_{dA}\lambda_{rd}i_{rq} = -$$

$$i_{sd} = i_{md} - i_{rd} = -$$

$$i_{rd} = + \text{ such that } i_{rd} > i_{md}$$

$$Q_r = v_{rq}i_{rd} = \omega_{dA}\lambda_{rd}i_{rd} = -$$

$$v_{rd} = 0$$

$$v_{rq} = \omega_{dA}\lambda_{rd} = -$$

(*Continued*)

$$\begin{bmatrix} v_A(t) \\ v_B(t) \\ v_C(t) \end{bmatrix} = \sqrt{\frac{2}{3}} \begin{bmatrix} \cos(\theta_{dA}) & -\sin(\theta_{dA}) \\ \cos\left(\theta_{dA} + \dfrac{4\pi}{3}\right) & -\sin\left(\theta_{dA} + \dfrac{4\pi}{3}\right) \\ \cos\left(\theta_{dA} + \dfrac{2\pi}{3}\right) & -\sin\left(\theta_{dA} + \dfrac{2\pi}{3}\right) \end{bmatrix} \begin{bmatrix} v_{rd} \\ v_{rq} \end{bmatrix}$$

Various space vectors are shown in Fig. 7-5.

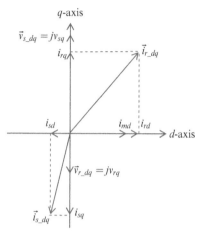

Fig. 7-5   Space vector diagram for Example 7-2.

7-2   DYNAMIC ANALYSIS OF DFIG

Equations for DFIG in terms of dq windings are the same as described in Chapter 3, where it is assumed that the rotor windings have the same number of turns as the stator windings, that is, $N_r = N_s$ However, for $n = (N_r/N_s)$, these equations can be rewritten, left as homework problems.

7-3   VECTOR CONTROL OF DFIG

In Chapter 5, vector control was described by aligning the d-axis with the rotor flux. However, in controlling DFIG, it is common to align the d-axis with the stator voltage vector since stator voltages are easy to measure [2]. With this choice of the reference frame, the d-axis stator current contributes to the real power P, and the q-axis stator current contributes to the reactive power of the DFIG. As discussed earlier, these are controlled by controlling the rotor currents i_{rd} and i_{rq}.

EXAMPLE 7-3

Consider a DFIG as a "test" machine, described in Chapter 1. Design the controller and show the output results.

Solution

A detailed controller design procedure and the results are on the accompanying website.

7-4   SUMMARY

Doubly fed induction generators (DFIGs) are used in harnessing wind energy. In this chapter, the principle of operation of doubly fed induction machines is described mathematically in order to apply vector control.

REFERENCES

1. N. Mohan, *Electric Machines and Drives*, Wiley, Hoboken, NJ, 2012. http://www.wiley.com/college/mohan.
2. T. Brekken, "A Novel Control Scheme for a Doubly-Fed Wind Generator under Unbalanced Grid Voltage Conditions," PhD thesis, University of Minnesota, July 2005.

PROBLEMS

7-1   A DFIG is operating in the motoring mode at a subsynchronous speed at a leading power factor (supplying Q_s from the grid).

Calculate the signs of various quantities in this mode of operation.

7-2 A DFIG is operating in the generator mode at a subsynchronous speed at a leading power factor (supplying Q_s to the grid). Calculate the signs of various quantities in this mode of operation.

7-3 Equations for DFIG in terms of dq windings are the same as described in Chapter 3, where it is assumed that the rotor windings have the same number of turns as the stator windings, that is, $N_r = N_s$. However, write these equations for $n = (N_r/N_s)$ and draw dq winding equivalent circuits

   (a) "Seen" from the stator-side.
   (b) "Seen" from the rotor-side.

8 Space Vector Pulse Width-Modulated (SV-PWM) Inverters

8-1 INTRODUCTION

In Chapter 5, we briefly discussed current-regulated pulse width-modulated (PWM) inverters using current-hysteresis control, in which the switching frequency f_s does not remain constant. The desired currents can also be supplied to the motor by calculating and then applying appropriate voltages, which can be generated based on the sinusoidal pulse-width-modulation principles discussed in basic courses in electric drives and power electronics [1]. However, the availability of digital signal processors in control of electric drives provides an opportunity to improve upon this sinusoidal pulse-width modulation by a procedure described in this chapter [2,3], which is termed space vector pulse-width modulation (SV-PWM). We will simulate such an inverter using Simulink for use in ac drives.

8-2 SYNTHESIS OF STATOR VOLTAGE SPACE VECTOR $\vec{v}_s^a$

In terms of the instantaneous stator phase voltages, the stator space voltage vector is

$$\vec{v}_s^a(t) = v_a(t)e^{j0} + v_b(t)e^{j2\pi/3} + v_c(t)e^{j4\pi/3}. \qquad (8-1)$$

In the circuit of Fig. 8-1, in terms of the inverter output voltages with respect to the negative dc bus and hypothetically assuming the stator neutral as a reference ground

Advanced Electric Drives: Analysis, Control, and Modeling Using MATLAB/Simulink®, First Edition. Ned Mohan.
© 2014 John Wiley & Sons, Inc. Published 2014 by John Wiley & Sons, Inc.

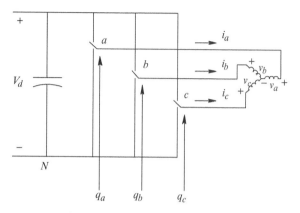

Fig. 8-1 Switch-mode inverter.

$$v_a = v_{aN} + v_N; \quad v_b = v_{bN} + v_N; \quad v_c = v_{cN} + v_N. \tag{8-2}$$

Substituting Eq. (8-2) into Eq. (8-1) and recognizing that

$$e^{j0} + e^{j2\pi/3} + e^{j4\pi/3} = 0, \tag{8-3}$$

the instantaneous stator voltage space vector can be written in terms of the inverter output voltages as

$$\vec{v}_s^a(t) = v_{aN}e^{j0} + v_{bN}e^{j2\pi/3} + v_{cN}e^{j4\pi/3}. \tag{8-4}$$

A switch in an inverter pole of Fig. 8-1 is in the "up" position if the pole switching function $q = 1$, otherwise in the "down" position if $q = 0$. In terms of the switching functions, the instantaneous voltage space vector can be written as

$$\vec{v}_s^a(t) = V_d(q_a e^{j0} + q_b e^{j2\pi/3} + q_c e^{j4\pi/3}). \tag{8-5}$$

With three poles, eight switch-status combinations are possible. In Eq. (8-5), the stator voltage vector $\vec{v}_s^a(t)$ can take on one of the following seven distinct instantaneous values, where in a digital representation, phase "a" represents the least significant digit and phase "c" the most significant digit (e.g., the resulting voltage vector due to the switch-status combination $\underset{(=3)}{011}$ is represented as $\vec{v}_3$):

$$\vec{v}_s^a(000) = \vec{v}_0 = 0$$
$$\vec{v}_s^a(001) = \vec{v}_1 = V_d e^{j0}$$
$$\vec{v}_s^a(010) = \vec{v}_2 = V_d e^{j2\pi/3}$$
$$\vec{v}_s^a(011) = \vec{v}_3 = V_d e^{j\pi/3}$$
$$\vec{v}_s^a(100) = \vec{v}_4 = V_d e^{j4\pi/3} \qquad (8\text{-}6)$$
$$\vec{v}_s^a(101) = \vec{v}_5 = V_d e^{j5\pi/3}$$
$$\vec{v}_s^a(110) = \vec{v}_6 = V_d e^{j\pi}$$
$$\vec{v}_s^a(111) = \vec{v}_7 = 0.$$

In Eq. (8-6), $\vec{v}_0$ and $\vec{v}_7$ are the zero vectors because of their values. The resulting instantaneous stator voltage vectors, which we will call the "basic vectors," are plotted in Fig. 8-2. The basic vectors form six sectors, as shown in Fig. 8-2.

The objective of the SV-PWM control of the inverter switches is to synthesize the desired reference stator voltage space vector in an optimum manner with the following objectives:

- A constant switching frequency f_s
- Smallest instantaneous deviation from its reference value
- Maximum utilization of the available dc-bus voltages

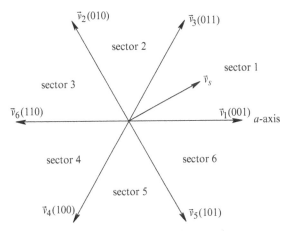

Fig. 8-2 Basic voltage vectors ($\vec{v}_0$ and $\vec{v}_7$ not shown).

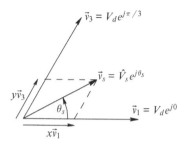

Fig. 8-3   Voltage vector in sector 1.

- Lowest ripple in the motor current, and
- Minimum switching loss in the inverter.

The above conditions are generally met if the average voltage vector is synthesized by means of the two instantaneous basic nonzero voltage vectors that form the sector (in which the average voltage vector to be synthesized lies) and both the zero voltage vectors, such that each transition causes change of only one switch status to minimize the inverter switching loss.

In the following analysis, we will focus on the average voltage vector in sector 1 with the aim of generalizing the discussion to all sectors. To synthesize an average voltage vector $\vec{v}_s^a$ ($=\hat{V}_s e^{j\theta_s}$) over a time period T_s in Fig. 8-3, the adjoining basic vectors $\vec{v}_1$ and $\vec{v}_3$ are applied for intervals xT_s and yT_s, respectively, and the zero vectors $\vec{v}_0$ and $\vec{v}_7$ are applied for a total duration of zT_s. In terms of the basic voltage vectors, the average voltage vector can be expressed as

$$\vec{v}_s^a = \frac{1}{T_s}[xT_s\vec{v}_1 + yT_s\vec{v}_3 + zT_s \cdot 0] \tag{8-7}$$

or

$$\vec{v}_s^a = x\vec{v}_1 + y\vec{v}_3, \tag{8-8}$$

where

$$x + y + z = 1. \tag{8-9}$$

In Eq. (8-8), expressing voltage vectors in terms of their amplitude and phase angles results in

$$\hat{V}_s e^{j\theta_s} = xV_d e^{j0} + yV_d e^{j\pi/3}. \qquad (8\text{-}10)$$

By equating real and imaginary terms on both sides of Eq. (8-10), we can solve for x and y (in terms the given values of $\hat{V}_s$, θ_s, and V_d) to synthesize the desired average space vector in sector 1 (see Problem 8-1).

Having determined the durations for the adjoining basic vectors and the two zero vectors, the next task is to relate the earlier discussion to the actual poles (a, b, and c). Note in Fig. 8-2 that in any sector, the adjoining basic vectors differ in one position; for example, in sector 1 with the basic vectors $\vec{v}_1(001)$ and $\vec{v}_3(011)$, only the pole "b" differs in the switch position. For sector 1, the switching pattern in Fig. 8-4 shows that pole-a is in "up" position during the sum of xT_s, yT_s, and z_7T_s intervals, and hence for the longest interval of the three poles. Next in the length of duration in the "up" position is pole-b for the sum of yT_s, and z_7T_s intervals. The smallest in the length of duration is pole-c for only z_7T_s interval. Each transition requires a change in switch state in only one of the poles, as shown in Fig. 8-4. Similar switching

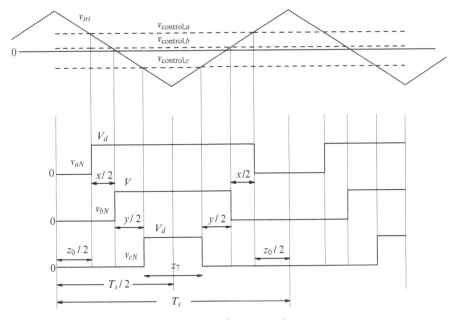

Fig. 8-4   Waveforms in sector 1; $z = z_0 + z_7$.

patterns for the three poles can be generated for any other sector (see Problem 8-2).

8-3 COMPUTER SIMULATION OF SV-PWM INVERTER

In computer simulations, for example, using Simulink, as well as in hardware implementation using rapid prototyping tools such as from DSPACE [4], the earlier described pulse-width modulation of the stator voltage space vector can be carried out by comparing control voltages with a triangular waveform signal at the switching frequency to generate switching functions. It is similar to the sinusoidal PWM approach only to the extent of comparing control voltages with a triangular waveform signal. However, in SV-PWM, the control voltages do not have a purely sinusoidal nature as those in the sinusoidal PWM.

In an induction machine with an isolated neutral, the three-phase voltages sum to zero (see Problem 8-3)

$$v_a(t) + v_b(t) + v_c(t) = 0. \tag{8-11}$$

To synthesize an average space vector $\vec{v}_s^a$ with phase components v_a, v_b, and v_c (the dc-bus voltage V_d is specified), the control voltages can be written in terms of the phase voltages as follows, expressed as a ratio of $\hat{V}_{tri}$ (the amplitude of the constant switching frequency triangular signal v_{tri} used for comparison with these control voltages):

$$\frac{v_{control,a}}{\hat{V}_{tri}} = \frac{v_a - v_k}{V_d / 2}$$

$$\frac{v_{control,b}}{\hat{V}_{tri}} = \frac{v_b - v_k}{V_d / 2} \tag{8-12}$$

$$\frac{v_{control,c}}{\hat{V}_{tri}} = \frac{v_c - v_k}{V_d / 2}.$$

where

$$v_k = \frac{\max(v_a, v_b, v_c) + \min(v_a, v_b, v_c)}{2}. \tag{8-13}$$

Deriving Eq. (8-13) is left as a homework problem (Problem 8-5).

EXAMPLE 8-1

In a three-phase inverter, the dc bus voltage $V_d = 700\,\text{V}$. Using the space vector modulation principles, calculate and plot the control voltages in steady state to synthesize a 60-Hz output with a line-line rms value of 460V. Assume that $\hat{V}_{\text{tri}} = 5\,\text{V}$ and the switching frequency $f_s = 10\,\text{kHz}$.

Solution

Fig. 8-5 shows the block diagram in Simulink, which is included on the accompanying website, to synthesize the ac output voltages. The results are plotted in Fig. 8-6.

8-4   LIMIT ON THE AMPLITUDE $\hat{V}_s$ OF THE STATOR VOLTAGE SPACE VECTOR $\vec{v}_s^a$

First, we will establish the absolute limit on the amplitude $\hat{V}_s$ of the average stator voltage space vector at various angles. The limit on the amplitude equals V_d (the dc-bus voltage) if the average voltage vector lies along a nonzero basic voltage vector. In between the basic vectors, the limit on the average voltage vector amplitude is that its tip can lie on the straight lines shown in Fig. 8-7, forming a hexagon (see Problem 8-6).

However, the maximum amplitude of the output voltage $\vec{v}_s^a$ should be limited to the circle within the hexagon in Fig. 8-7 to prevent distortion in the resulting currents. This can be easily concluded from the fact that in a balanced sinusoidal steady state, the voltage vector $\vec{v}_s^a$ rotates at the synchronous speed with its constant amplitude. At its maximum amplitude,

$$\vec{v}_{s,\text{max}}^a(t) = \hat{V}_{s,\text{max}} e^{j\omega_{\text{syn}} t}. \tag{8-14}$$

Therefore, the maximum value that $\hat{V}_s$ can attain is

$$\hat{V}_{s,\text{max}} = V_d \cos\left(\frac{60^0}{2}\right) = \frac{\sqrt{3}}{2} V_d. \tag{8-15}$$

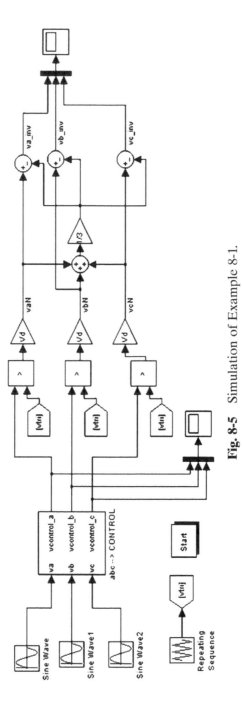

Fig. 8-5 Simulation of Example 8-1.

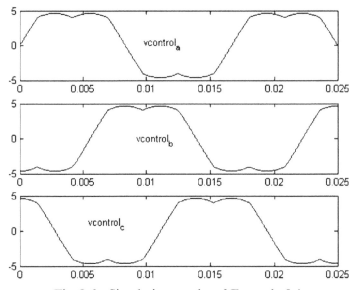

Fig. 8-6   Simulation results of Example 8-1.

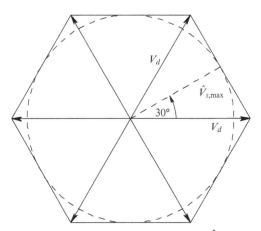

Fig. 8-7   Limit on amplitude $\hat{V}_s$.

From Eq. (8-15), the corresponding limits on the phase voltage and the line–line voltages are as follows:

$$\hat{V}_{\text{phase,max}} = \frac{2}{3}\hat{V}_{s,\text{max}} = \frac{V_d}{\sqrt{3}} \qquad (8\text{-}16)$$

and

$$V_{LL,\text{max}}(rms) = \sqrt{3}\frac{\hat{V}_{\text{phase,max}}}{\sqrt{2}} = \frac{V_d}{\sqrt{2}} = 0.707V_d. \qquad (8\text{-}17)$$

The sinusoidal pulse-width modulation in the linear range discussed in the previous course on electric drives and power electronics results in a maximum voltage

$$V_{LL,\text{max}}(rms) = \frac{\sqrt{3}}{2\sqrt{2}}V_d = 0.612V_d \quad \text{(sinusoidal PWM).} \qquad (8\text{-}18)$$

Comparison of Eq. (8-17) and Eq. (8-18) shows that the SV-PWM discussed in this chapter better utilizes the dc bus voltage and results in a higher limit on the available output voltage by a factor of $(2/\sqrt{3})$, or by approximately 15% higher, compared with the sinusoidal PWM.

SUMMARY

In this chapter, an approach called SV-PWM is discussed, which is better than the sinusoidal PWM approach in utilizing the available dc-bus voltage. Its modeling using Simulink is described.

REFERENCES

1. N. Mohan, *Electric Machines and Drives: A First Course*, Wiley, Hoboken, NJ, 2011. http://www.wiley.com/college/mohan.
2. H.W. van der Broek et al., "Analysis and Realization of a Pulse Width Modulator Based on Voltage Space Vectors," IEEE Industry Applications Society Proceedings, 1986, pp. 244–251.

3. J. Holtz, "Pulse Width Modulation for Electric Power Converters," chapter 4 in *Power Electronics and Variable Frequency Drives*, ed. B.K. Bose, IEEE Press, New York, 1997.

4. http://www.dspace.de.

PROBLEMS

8-1 In a converter, $V_d = 700$ V. To synthesize an average stator voltage vector $\vec{v}_s^a = 563.38e^{j0.44}$ V, calculate $x, y,$ and z.

8-2 Repeat if $\vec{v}_s^a = 563.38e^{j2.53}$ V. Plot results similar to those in Fig. 8-4.

8-3 Show that in an induction machine with isolated neutral, at any instant of time, $v_a(t) + v_b(t) + v_c(t) = 0$.

8-4 Given that $\vec{v}_s^a = 563.38e^{j0.44}$ V, calculate the phase voltage components.

8-5 Derive Eq. (8-12).

8-6 Derive that the maximum limit on the amplitude of the space vector forms the hexagonal trajectory shown in Fig. 8-7.

9 Direct Torque Control (DTC) and Encoderless Operation of Induction Motor Drives

9-1 INTRODUCTION

Unlike vector-control techniques described in previous chapters, in the direct-control (DTC) scheme, no dq-axis transformation is needed, and the electromagnetic torque and the stator flux are estimated and directly controlled by applying the appropriate stator voltage vector [1–3]. It is possible to estimate the rotor speed, thus eliminating the need for rotor speed encoder.

9-2 SYSTEM OVERVIEW

Figure 9-1 shows the block diagram of the overall system, which includes the speed and the torque feedback loops, without a speed encoder. The estimated speed $\omega_{mech,est}$ is subtracted from the reference (desired) speed ω_{mech}^*, and the error between the two acts on a PI-controller to generate the torque reference signal T_{em}^*. The estimated speed generates the reference signal for the stator flux linkage $\hat{\lambda}_s^*$ (thus allowing flux weakening for extended range of speed operation), which is compared with the estimated stator flux linkage $\hat{\lambda}_{s,est}$. The errors in the electromagnetic torque and the stator flux, combined with the angular position $\angle\theta_s$ of the stator flux linkage space vector, determine the stator voltage space vector $\vec{v}_s$ that is applied to the motor during each sampling interval ΔT, for example, equal to $25\,\mu s$.

Advanced Electric Drives: Analysis, Control, and Modeling Using MATLAB/Simulink®, First Edition. Ned Mohan.
© 2014 John Wiley & Sons, Inc. Published 2014 by John Wiley & Sons, Inc.

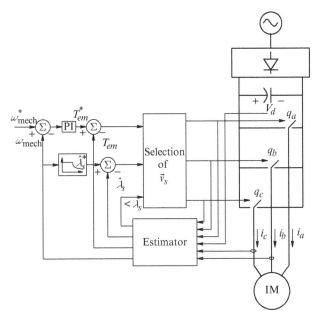

Fig. 9-1   Block diagram of DTC.

Estimating the electromagnetic torque and the stator flux linkage vector requires measuring the stator currents and the stator phase voltages—the latter, as shown in Fig. 9-1, are indirectly calculated by measuring the dc-bus voltage and knowing within the digital controller the status of the inverter switches.

9-3   PRINCIPLE OF ENCODERLESS DTC OPERATION

Prior to detailed derivations, we can enumerate the various steps in the estimator block of Fig. 9-1 as follows, where all space vectors are implicitly expressed in electrical radians with respect to the stator a-axis as the reference axis (unless explicitly mentioned otherwise):

1. From the measured stator voltages and currents, calculate the stator flux linkage space vector $\vec{\lambda}_s$:

$$\vec{\lambda}_s(t) = \vec{\lambda}_s(t - \Delta T) + \int_{t-\Delta T}^{t} (\vec{v}_s - R_s \vec{i}_s) \cdot d\tau = \hat{\lambda}_s e^{j\theta_s}.$$

2. From $\vec{\lambda}_s$ and $\vec{i}_s$, calculate the rotor flux space vector $\vec{\lambda}_r$ and hence the speed of the rotor flux linkage vector, where ΔT_ω is a sampling time for speed calculation:

$$\vec{\lambda}_r = \frac{L_r}{L_m}(\vec{\lambda}_s - \sigma L_s \vec{i}_s) = \hat{\lambda}_r e^{j\theta_r} \quad \text{and} \quad \omega_r = \frac{d}{dt}\theta_r = \frac{\theta_r(t) - \theta_r(t - \Delta T_\omega)}{\Delta T_\omega}.$$

3. From $\vec{\lambda}_s$ and $\vec{i}_s$, calculate the estimated electromagnetic torque T_{em}:

$$T_{em} = \left(\frac{2}{3}\right)\frac{P}{2}\text{Im}(\vec{\lambda}_s^{\text{conj}}\vec{i}_s).$$

4. From $\vec{\lambda}_r$ and $T_{em,est}$, estimate the slip speed ω_{slip} and the rotor speed ω_m:

$$\omega_{slip} = \frac{2}{p}\left(\frac{3}{2}R_r\frac{T_{em}}{\hat{\lambda}_r^2}\right) \quad \text{and} \quad \omega_m = \omega_r - \omega_{slip}.$$

In the stator voltage selection block of Fig. 9-1, an appropriate stator voltage vector is calculated to be applied for the next sampling interval ΔT based on the errors in the torque and the stator flux, in order to keep them within a hysteretic band.

9-4   CALCULATION OF $\vec{\lambda}_s$, $\vec{\lambda}_r$, T_{em}, AND ω_m

9-4-1   Calculation of the Stator Flux $\vec{\lambda}_s$

The stator voltage equation with the stator a-axis as the reference is

$$\vec{v}_s = R_s\vec{i}_s + \frac{d}{dt}\vec{\lambda}_s. \tag{9-1}$$

From Eq. (9-1), the stator flux linkage space vector at time t can be calculated in terms of the flux linkage at the previous sampling time as

$$\vec{\lambda}_s(t) = \vec{\lambda}_s(t - \Delta T) + \int_{t-\Delta T}^{t}(\vec{v}_s - R_s\vec{i}_s)\cdot d\tau = \hat{\lambda}_s e^{j\theta_s}, \tag{9-2}$$

where τ is the variable of integration, the applied stator voltage remains constant during the sampling interval ΔT, and the stator current value is that measured at the previous time step.

9-4-2   Calculation of the Rotor Flux $\vec{\lambda}_r$

From Chapter 3,

$$\vec{\lambda}_s = L_s \vec{i}_s + L_m \vec{i}_r \tag{9-3}$$

and

$$\vec{\lambda}_r = L_r \vec{i}_r + L_m \vec{i}_s. \tag{9-4}$$

Calculating $\vec{i}_r$ from Eq. (9-3),

$$\vec{i}_r = \frac{\vec{\lambda}_s}{L_m} - \frac{L_s}{L_m} \vec{i}_s, \tag{9-5}$$

and substituting it into Eq. (9-4),

$$
\begin{aligned}
\vec{\lambda}_r &= \frac{L_r}{L_m} \vec{\lambda}_s - \frac{L_s L_r}{L_m} \vec{i}_s + L_m \vec{i}_s \\
&= \frac{L_r}{L_m} \left[\vec{\lambda}_s - L_s \vec{i}_s \underbrace{\left(1 - \frac{L_m^2}{L_s L_r} \right)}_{(=\sigma)} \right],
\end{aligned}
\tag{9-6}
$$

where the leakage factor σ is defined as (similar to Eq. 5-15 in Chapter 5)

$$\sigma = 1 - \frac{L_m^2}{L_s L_r}. \tag{9-7}$$

Therefore, the rotor flux linkage space vector in Eq. (9-6) can be written as

$$\vec{\lambda}_r = \frac{L_r}{L_m} (\vec{\lambda}_s - \sigma L_s \vec{i}_s) = \hat{\lambda}_r e^{j\theta_r}. \tag{9-8}$$

We should note that similar to Eq. (9-2), for the stator flux linkage vector, the rotor flux linkage space vector can be expressed as follows, recognizing that the rotor voltage in a squirrel-cage rotor is zero

$$\vec{\lambda}_r^A(t) = \vec{\lambda}_r^A(t - \Delta T) + \int_{t-\Delta T}^{t} (-R_r \vec{i}_r^A) \cdot d\tau = \hat{\lambda}_r e^{j\theta_r^A}, \tag{9-9}$$

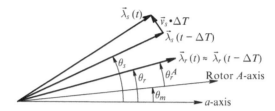

Fig. 9-2   Changing the position of stator flux-linkage vector.

where the space vectors and angles (in electrical radians) are expressed with respect to the rotor A-axis shown in Fig. 9-2. The above equation shows that the rotor flux changes very slowly with time (in amplitude and in phase angle θ_r^A with respect to the rotor A-axis) only due to a small voltage drop across the rotor resistance.

9-4-3   Calculation of the Electromagnetic Torque T_{em}

The electromagnetic torque developed by the motor can be estimated in terms of the stator flux and the stator current, or in terms of the stator flux and the rotor flux. We will derive both expressions in Appendix 9-A; however, the final expressions that we need are given below.

Torque depends on the magnitude of the stator and the rotor fluxes, and the angle between the two space vectors. As derived in Appendix 9-A, in terms of the machine leakage inductance L_σ (also defined in Appendix 9-A)

$$T_{em} = \left(\frac{2}{3}\right)\frac{p}{2}\frac{L_m}{L_\sigma^2}\hat{\lambda}_s\hat{\lambda}_r \sin\theta_{sr}, \tag{9-10a}$$

where

$$\theta_{sr} = \theta_s - \theta_r. \tag{9-10b}$$

The angles in Eq. (9-10) are expressed in electrical radians with respect the stator a-axis, as shown in Fig. 9-2.

Equation (9-2) and Fig. 9-2 show that the torque can be controlled quickly by rapidly changing the position of the stator flux linkage space vector (i.e., θ_s, hence θ_{sr}) by applying an appropriate voltage space vector during the sampling interval ΔT, while the rotor flux space vector position $\theta_r(=\theta_r^A+\theta_m)$ changes relatively slowly. Thus, in accordance with Eq. (9-10a), a change in θ_{sr} results in the desired change in torque.

For torque estimation, it is better to use the expression below (derived in Appendix 9-A) in terms of the estimated stator flux linkage and the measured stator currents,

$$T_{em}=\left(\frac{2}{3}\right)\frac{p}{2}\,\mathrm{Im}(\vec{\lambda}_s^{\,\mathrm{conj}}\vec{i}_s),\qquad(9\text{-}11)$$

which, unlike the expression in Eq. (9-10a), does not depend on the rotor flux linkage (note that the rotor flux linkage in Eq. (9-8) depends on correct estimates of L_s, L_r, and L_m).

9-4-4 Calculation of the Rotor Speed ω_m

A much slower sampling rate with a sampling interval ΔT_ω, for example, equal to 1 ms, may be used for estimating the rotor speed. The speed of the rotor flux in electrical radians per second (rad/s) is calculated from the phase angle of the rotor flux space vector in Eq. (9-8) as follows:

$$\omega_r=\frac{d}{dt}\theta_r=\frac{\theta_r(t)-\theta_r(t-\Delta T_\omega)}{\Delta T_\omega}.\qquad(9\text{-}12)$$

The slip speed is calculated as follows: In Chapter 5, the torque and the speed expressions are given by Eq. (5-7) and Eq. (5-5), where in the motor model, the d-axis is aligned with the rotor flux linkage space vector. These equations are repeated below:

$$T_{em}=\frac{p}{2}\lambda_{rd}\left(\frac{L_m}{L_r}i_{sq}\right)\qquad(9\text{-}13)$$

and

$$\omega_{\mathrm{slip}}=R_r\frac{1}{\lambda_{rd}}\left(\frac{L_m}{L_r}i_{sq}\right),\qquad(9\text{-}14)$$

where ω_{slip} is the slip speed, the same as ω_{dA} in Eq. (5-5) of Chapter 5. Calculating i_{sq} from Eq. (9-13) and substituting it into Eq. (9-14) (and recognizing that in the model with the d-axis aligned with the rotor flux linkage, $\lambda_{rd} = \sqrt{2/3}\hat{\lambda}_r$), the slip speed in electrical radians per second is

$$\omega_{slip} = \frac{2}{p}\left(\frac{3}{2}R_r\frac{T_{em}}{\hat{\lambda}_r^2}\right). \tag{9-15}$$

Therefore, the rotor speed can be estimated from Eq. (9-12) and Eq. (9-15) as

$$\omega_m = \omega_r - \omega_{slip}, \tag{9-16}$$

where all speeds are in electrical radians per second. In a multipole machine with $p \geq 2$,

$$\omega_{mech} = (2/p)\omega_m. \tag{9-17}$$

9-5 CALCULATION OF THE STATOR VOLTAGE SPACE VECTOR

A common technique in DTC is to control the torque and the stator flux amplitude with a hysteretic band around their desired values. Therefore, at a sampling time (with a sampling interval of ΔT), the decision to change the voltage space vector is implemented only if the torque and/or the stator flux amplitude are outside their range. Selection of the new voltage vector depends on the signs of the torque and the flux errors and the sector in which the stator flux linkage vector lies, as explained later.

The plane of the stator voltage space vector is divided into six sectors, as shown in Fig. 9-3. We should note that these sectors are different than those defined for the stator voltage space vector-PWM in Chapter 8. The central vectors for each sector, which lie in the middle of a sector, are the basic inverter vectors, as shown in Fig. 9-3.

The choice of the voltage space vector for sector 1 is explained later with the help of Fig. 9-4 and Eq. (9-10). Assuming that the stator flux linkage space vector is along the central vector, the roles of various voltage vectors can be tabulated in Table 9-1.

There are some additional observations: The voltage vectors would have the same effects as tabulated earlier, provided the stator flux-linkage space vector is anywhere in sector 1. The use of voltage vectors

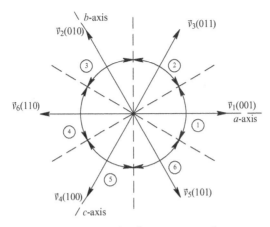

Fig. 9-3   Inverter basic vectors and sectors.

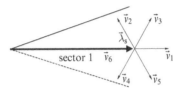

Fig. 9-4   Stator voltage vector selection in sector 1.

TABLE 9-1   Effect of Voltage Vector on the
Stator Flux-Linkage Vector in Sector 1

$\vec{v}_s$	T_{em}	$\hat{\lambda}_s$
$\vec{v}_3$	Increase	Increase
$\vec{v}_2$	Increase	Decrease
$\vec{v}_4$	Decrease	Decrease
$\vec{v}_5$	Decrease	Increase

$\vec{v}_1$ and $\vec{v}_6$ is avoided because their effect depends on where the stator flux-linkage vector is in sector 1. A similar table can be generated for all other sectors.

Use of zero vectors $\vec{v}_0(000)$ and $\vec{v}_7(111)$ results in the stator flux linkage vector essentially unchanged in amplitude and in the angular position θ_s. In the torque expression of Eq. (9-10b), for small values of θ_{sr} in electrical radians,

$$\sin\theta_{sr} \approx (\theta_s - \theta_r). \tag{9-18}$$

With a zero voltage vector applied, assuming that the amplitudes of the stator and the rotor flux linkage vectors remain constant,

$$T_{em} \simeq k(\theta_s - \theta_r), \tag{9-19}$$

where k is a constant. With the zero voltage vector applied, the position of the stator flux-linkage vector remains essentially constant, thus $\Delta\theta_s \simeq 0$. Similarly, the position of the rotor flux-linkage vector, with respect to the rotor A-axis, remains essentially constant, that is, $\Delta\theta_r^A \simeq 0$. However, as can be observed from Fig. 9-2, $\Delta\theta_r = \Delta\theta_m + \Delta\theta_r^A$. Therefore the position of the rotor flux-linkage vector changes, albeit slowly, and the change in torque in Eq. (9-19) can be expressed as

$$\Delta T_{em} \simeq -k(\Delta\theta_m) \quad \text{(with zero voltage vector applied)}. \tag{9-20}$$

Equation (9-20) shows that applying a zero stator voltage space vector causes change in torque in a direction opposite to that of ω_m. Therefore, with the rotor rotating in a positive (counter-clockwise) direction, for example, it may be preferable to apply a zero voltage vector to decrease torque in order to keep it within a hysteretic band.

In literature, there is no uniformity on the logic of space vector selection to keep the stator flux amplitude and the electromagnetic torque within their respective hysteretic bands. One choice of space vectors is illustrated by means of Example 9-1.

EXAMPLE 9-1

The "test" induction motor described in Chapter 1 is operated using encoderless DTC for speed control, as described in a file Ex9_1.pdf (which can be printed) on the website accompanying this textbook. Model this system using Simulink and plot the desired results.

Solution

Various subsystems and the simulation results are included on the website associated with this textbook.

9-6 DIRECT TORQUE CONTROL USING *dq*-AXES

It is possible to perform the same type of control by aligning the *d*-axis with the stator flux-linkage vector. The amplitude of the stator flux-linkage vector is controlled by applying v_{sd} along the *d*-axis, and the torque is controlled by applying v_{sq} along the *q*-axis. The advantage of this type of control over the hysteretic control described earlier is that it results in a constant switching frequency. This is described by Example 9-2 in accompanying website.

9-7 SUMMARY

This chapter discusses the direct torque control (DTC) scheme, where, unlike the vector control, no *dq*-axis transformation is needed and the electromagnetic torque and the stator flux are estimated and directly controlled by applying the appropriate stator voltage vector. It is possible to estimate the rotor speed, thus eliminating the need for rotor speed encoder.

REFERENCES

1. M. Depenbrock, "Direct Self Control (DSC) of Inverter-Fed Induction Machines," IEEE Transactions on Power Electronics, 1988, pp. 420–429.
2. I. Takahashi and T. Noguchi, "A New Quick Response and High Efficiency Strategy of an Induction Motor," IEEE Transactions on Industry Applications, vol. 22, no. 7, 1986, pp. 820–827.
3. P. Tiitinen and M. Surendra, "The Next Generation Motor Control Method, DTC Direct Torque Control," Proceedings of the International Conference on Power Electronics, Drives and Energy Systems, PEDES'96 New Delhi (India), pp. 37–43.

PROBLEMS

9-1 Using the parameters of the "test" induction machine described in Chapter 1, show that it is much faster to change electromagnetic torque by changing the position of the stator flux-linkage vector, rather than by changing its amplitude. Assume that the machine is operating under rated conditions.

9-2 Obtain the stator voltage vectors needed in other sectors, similar to what has been done in Table 9-1 for sector 1.

9-3 Assuming that the "test" machine is operating under the rated conditions, compute the effect of applying a zero voltage space vector on the flux linkage space vectors and on the electromagnetic torque produced.

9-4 In the system of Example 9-1, how can the modeling be simplified if the speed is never required to reverse?

9-5 Experiment with other schemes for selecting voltage space vector and compare results with that in Example 9-1.

9-6 In Example 9-1, include the field-weakening mode of operation.

APPENDIX 9-A

Derivation of Torque Expressions

The electromagnetic torque in terms of the stator flux and the stator current can be expressed as follows:

$$T_{em} = \frac{2}{3}\frac{p}{2}\,\text{Im}(\vec{\lambda}_s^{conj}\vec{i}_s) \tag{9A-1}$$

To derive the above expression, it is easiest to assume it be correct and to substitute the components to prove it. Taking the complex conjugate on both sides of Eq. 9-3,

$$\vec{\lambda}_s^{conj} = L_s\vec{i}_s^{conj} + L_m\vec{i}_r^{conj} \tag{9A-2}$$

Substituting in Eq. 9A-1,

$$T_{em} = \frac{2}{3}\frac{p}{2}\{L_s\underbrace{\text{Im}(\vec{i}_s^{conj}\vec{i}_s)}_{(=0)} + L_m\,\text{Im}(\vec{i}_r^{conj}\vec{i}_s)\} = \frac{2}{3}\frac{p}{2}L_m\,\text{Im}(\vec{i}_r^{conj}\vec{i}_s) \tag{9A-3}$$

Even though the *dq* transformation is not used in DTC, we can make use of *dq* transformations to prove our expressions. Therefore, in terms of an arbitrary *dq* reference set and the corresponding components substituted in Eq. 9A-3,

$$T_{em} = \frac{2}{3}\frac{p}{2}L_m \, \text{Im}\{\sqrt{\frac{3}{2}}(i_{rd} - ji_{rq})\sqrt{\frac{3}{2}}(i_{sd} + ji_{sq})\} = \frac{p}{2}L_m(i_{sq}i_{rd} - i_{sd}i_{rq})$$

$$(9A\text{-}4)$$

which is identical to Eq. 3-47 of Chapter 3, thus proving the torque expression of Eq. 9A-1 to be correct.

Another torque expression, which we will not use directly but which is the basis on which the selection of the stator voltage vector is made, is as follows:

$$T_{em} = \frac{2}{3}\frac{p}{2}\frac{L_m}{L_\sigma^2} \, \text{Im}(\vec{\lambda}_s \vec{\lambda}_r^{conj})$$

$$(9A\text{-}5)$$

where the machine leakage inductance is defined as

$$L_\sigma = \sqrt{L_s L_r - L_m^2}$$

$$(9A\text{-}6)$$

Again assuming the above expression in Eq. 9A-5 to be correct and substituting the expressions for the fluxes from Eqs. 9-3 and 9-4,

$$T_{em} = \frac{2}{3}\frac{p}{2}\frac{L_m}{L_\sigma^2} \, \text{Im}\{(L_s\vec{i}_s + L_m\vec{i}_r)(L_r\vec{i}_r^{conj} + L_m\vec{i}_s^{conj})\}$$

$$= \frac{2}{3}\frac{p}{2}\frac{L_m}{L_\sigma^2} L_s L_r \, \text{Im}(\vec{i}_s\vec{i}_r^{conj}) + \frac{2}{3}\frac{p}{2}\frac{L_m}{L_\sigma^2} L_m^2 \, \text{Im}(\vec{i}_r\vec{i}_s^{conj}).$$

$$(9A\text{-}7)$$

Note that $\text{Im}(\vec{i}_r\vec{i}_s^{conj}) = -\text{Im}(\vec{i}_s\vec{i}_r^{conj})$. Therefore, in Eq. 9A-7,

$$T_{em} = \frac{2}{3}\frac{p}{2}\frac{L_m}{L_\sigma^2}\underbrace{(L_s L_r - L_m^2)}_{(=L_\sigma^2)}\text{Im}(\vec{i}_s\vec{i}_r^{conj}) = \frac{2}{3}\frac{p}{2}L_m \, \text{Im}(\vec{i}_s\vec{i}_r^{conj})$$

$$= \frac{p}{2}L_m \, \text{Im}\{(i_{sd} + ji_{sq})(i_{rd} - ji_{rq})\}$$

$$(9A\text{-}8)$$

$$= \frac{p}{2}L_m(i_{sq}i_{rd} - i_{sd}i_{rq}),$$

which is identical to Eq. 3-47 of Chapter 3, thus proving the torque expression of Eq. 9A-5 to be correct.

In Eq. 9A-5, expressing flux linkages in their polar form,

$$
\begin{aligned}
T_{em} &= \frac{2}{3}\frac{p}{2}\frac{L_m}{L_\sigma^2}\mathrm{Im}(\hat{\lambda}_s e^{j\theta_s}\cdot\hat{\lambda}_r e^{-j\theta_r}) = \frac{2}{3}\frac{p}{2}\frac{L_m}{L_\sigma^2}\hat{\lambda}_s\hat{\lambda}_r\mathrm{Im}(e^{j\theta_{sr}}) \\
&= \frac{2}{3}\frac{p}{2}\frac{L_m}{L_\sigma^2}\hat{\lambda}_s\hat{\lambda}_r\sin\theta_{sr}
\end{aligned}
\tag{9A-9}
$$

where

$$
\theta_{sr} = \theta_s - \theta_r
\tag{9A-10}
$$

is the angle between the two flux-linkage space vectors.

10 Vector Control of Permanent-Magnet Synchronous Motor Drives

10-1 INTRODUCTION

In the previous course [1], we looked at permanent-magnet synchronous motor drives, also known as "brushless-dc motor" drives in steady state, where without the help of dq analysis, it was not possible to discuss dynamic control of such drives. In this chapter, we will make use of the dq-analysis of induction machines, which is easily extended to analyze and control synchronous machines.

10-2 d-q ANALYSIS OF PERMANENT MAGNET (NONSALIENT-POLE) SYNCHRONOUS MACHINES

In synchronous motors with surface-mounted permanent magnets, the rotor can be considered magnetically round (non-salient) that has the same reluctance along any axis through the center of the machine. A simplified representation of the rotor magnets is shown in Fig. 10-1a. The three-phase stator windings are sinusoidally distributed in space, like in an induction machine, with N_s number of turns per phase.

In Fig. 10-1b, d-axis is always aligned with the rotor magnetic axis, with the q-axis 90° ahead in the direction of rotation, assumed to be counter-clockwise. The stator three-phase windings are represented by equivalent d- and q-axis windings; each winding has $\sqrt{3/2}N_s$ turns, which are sinusoidally distributed.

Advanced Electric Drives: Analysis, Control, and Modeling Using MATLAB/Simulink®, First Edition. Ned Mohan.

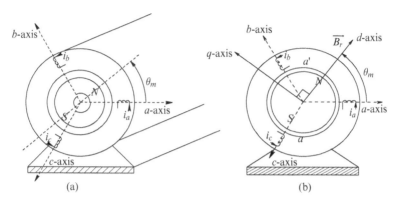

Fig. 10-1   Permanent-magnet synchronous machine (shown with $p = 2$).

10-2-1   Flux Linkages

The stator d and q winding flux linkages can be expressed as follows:

$$\lambda_{sd} = L_s i_{sd} + \lambda_{fd} \tag{10-1}$$

and

$$\lambda_{sq} = L_s i_{sq}, \tag{10-2}$$

where in Eq. (10-1) and Eq. (10-2), $L_s = L_{\ell s} + L_m$, and λ_{fd} is the flux linkage of the stator d winding due to flux produced by the rotor magnets (recognizing that the d-axis is always aligned with the rotor magnetic axis).

10-2-2   Stator dq Winding Voltages

Using Eq. (3-28) and Eq. (3-29), developed for induction machines in Chapter 3, in dq windings,

$$v_{sd} = R_s i_{sd} + \frac{d}{dt}\lambda_{sd} - \omega_m \lambda_{sq} \tag{10-3}$$

and

$$v_{sq} = R_s i_{sq} + \frac{d}{dt}\lambda_{sq} + \omega_m \lambda_{sd}, \tag{10-4}$$

where the speed of the equivalent *dq* windings is $w_d = w_m$ (in electrical rad/s) in order to keep the *d*-axis always aligned with the rotor magnetic axis [2]. The speed w_m is related to the actual rotor speed w_{mech} as

$$\omega_m = \frac{p}{2}\omega_{mech}.$$ (10-5)

10-2-3   Electromagnetic Torque

Using the analysis for induction machines in Chapter 3 and Eq. (3-46) and Eq. (3-47), we can derive the following equation (see Problem 3-9a in Chapter 3), which is also valid for synchronous machines:

$$T_{em} = \frac{p}{2}(\lambda_{sd}i_{sq} - \lambda_{sq}i_{sd}).$$ (10-6)

Substituting for flux linkages in the above equation for a nonsalient-pole machine,

$$T_{em} = \frac{p}{2}[(L_s i_{sd} + \lambda_{fd})i_{sq} - L_s i_{sq}i_{sd}] = \frac{p}{2}\lambda_{fd}i_{sq} \quad \text{(nonsalient)}.$$ (10-7)

10-2-4   Electrodynamics

The acceleration is determined by the difference of the electromagnetic torque and the load torque (including friction torque) acting on J_{eq}, the combined inertia of the load and the motor:

$$\frac{d}{dt}\omega_{mech} = \frac{T_{em} - T_L}{J_{eq}},$$ (10-8)

where w_{mech} is in rad/s and is related to w_m as shown in Eq. (10-5).

10-2-5   Relationship between the *dq* Circuits and the Per-Phase Phasor-Domain Equivalent Circuit in Balanced Sinusoidal Steady State

In this section, we will see that under a balanced sinusoidal steady state condition, the two *dq* winding equivalent circuits combine to result in the per-phase equivalent circuit of a synchronous machine that we have derived in the previous course. Note that in a synchronous motor used in a "brush-less dc" drive, the synchronous speed equals the rotor speed on an instantaneous basis, therefore our choice of $w_d = w_m$ also results in $w_d = w_m = w_{syn}$. Under a balanced sinusoidal steady-state condition, dq winding quantities are dc and their time derivatives are zero. In Eq. (10-3) and Eq. (10-4), for stator voltages, substituting flux linkages from Eq. (10-1) and Eq. (10-2) results in

$$v_{sd} = R_s i_{sd} - w_m L_s i_{sq} \tag{10-9}$$

and

$$v_{sq} = R_s i_{sq} + w_m L_s i_{sd} + w_m \lambda_{fd}. \tag{10-10}$$

Multiplying both sides of Eq. (10-10) by (j) and adding to Eq. (10-9) (and multiplying both sides of the resulting equation by $\sqrt{3/2}$) leads to the following space vector equation, with the d-axis as the reference axis:

$$\vec{v}_s = R_s \vec{i}_s + j w_m L_s \vec{i}_s + \underbrace{j\sqrt{3/2} w_m \lambda_{fd}}_{\vec{e}_{fs}}, \tag{10-11}$$

noting that $\vec{v}_s = \sqrt{3/2}(v_{sd} + j v_{sq})$ and so on. Dividing both sides of the above space vector equation by 3/2, we obtain the following phasor equation for phase a in a balanced sinusoidal steady state:

$$\bar{V}_a = R_s \bar{I}_a + j w_m L_s \bar{I}_a + \underbrace{j w_m \sqrt{\frac{2}{3}} \lambda_{fd}}_{\bar{E}_{fa}}. \tag{10-12}$$

The above equation corresponds to the per-phase equivalent circuit of Fig. 10-2 that was derived in the previous course under a balanced sinusoidal steady-state condition.

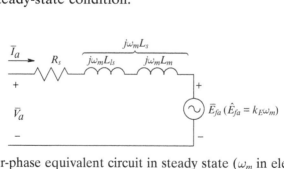

Fig. 10-2   Per-phase equivalent circuit in steady state (w_m in electrical rad/s).

Relationship between k_E and λ_{fd}
From Eq. (10-12),

$$\hat{E}_{fa} = \underbrace{\sqrt{\frac{2}{3}}\lambda_{fd}}_{k_E}\,\omega_m = k_E\omega_m, \qquad (10\text{-}13)$$

Therefore,

$$k_E = \sqrt{\frac{2}{3}}\lambda_{fd}. \qquad (10\text{-}14)$$

10-2-6   *dq*-Based Dynamic Controller for "Brushless DC" Drives

In the previous course, in the absence of the *dq* analysis, a hysteretic converter was used, where the switching frequency does not remain constant. In this section, we will see that it is possible to use a converter with a constant switching frequency. The block diagram of such a control system is shown in Fig. 10-3. In Eq. (10-3) and Eq. (10-4), using the flux linkages of Eq. (10-1) and Eq. (10-2), voltages can be expressed as follows, recognizing that the time-derivative of the rotor-produced flux λ_{fd} is zero:

$$v_{sd} = R_s i_{sd} + L_s \frac{d}{dt} i_{sd} + \underbrace{(-\omega_m L_s i_{sq})}_{\text{comp}_d} \qquad (10\text{-}15)$$

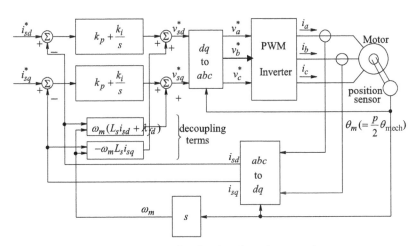

Fig. 10-3   Controller in the *dq* reference frame.

and

$$v_{sq} = R_s i_{sq} + L_s \frac{d}{dt} i_{sq} + \underbrace{\omega_m (L_s i_{sd} + \lambda_{fd})}_{\text{comp}_q}. \tag{10-16}$$

In Fig. 10-3, the PI controllers in both channels are designed assuming that the inverter is ideal and the compensation (decoupling) terms in Eq. (10-15) and Eq. (10-16) are utilized, to result in the desired phase margin at the chosen open-loop crossover frequency.

Flux Weakening   In the normal speed range below the rated speed, the reference for the d winding current is kept zero ($i_{ds} = 0$). Beyond the rated speed, a negative current in the d winding causes flux weakening (a phenomenon similar to that in brush-type dc machines and induction machines), thus keeping the back-emf from exceeding the rated voltage of the motor. A negative value of i_{sd} in Eq. (10-10) causes v_{sq} to decrease.

To operate synchronous machines with surface-mounted permanent magnets at above the rated speed requires a substantial negative d winding current to keep the terminal voltage from exceeding its rated value. Note that the total current into the stator cannot exceed its rated value in steady state. Therefore, the higher the magnitude of the d winding current, the lower the magnitude of the q winding current has to be, since

$$\sqrt{\left|i_{sd}\right|^2 + \left|i_{sq}\right|^2} \leq \hat{I}_{dq,\text{rated}} \left(= \sqrt{\frac{3}{2}} \hat{I}_{a,\text{rated}}\right). \tag{10-17}$$

EXAMPLE 10-1

For analyzing performance of the dynamic control procedure, a motor from a commercial vendor catalog [3] is selected, whose specifications are as follows:

Nameplate Data

Continuous Stall Torque: 3.2 Nm
Continuous Current: 8.74 A

Peak Torque: 12.8 Nm
Peak Current: 31.5 A
Rated Voltage: 200 V
Rated Speed: 6000 rpm
Phases: 3
Number of Poles: 4

Per-Phase Motor Circuit Parameters

$R_s = 0.416\,\Omega$

$L_s = 1.365\,\text{mH}$

Voltage Constant k_E (as in Eq. 10-13 and Fig. 10-2):   0.0957 V/ (electrical rad/s)

The total equivalent inertia of the system (motor–load combination) is

$$J_{eq} = 3.4 \times 10^{-4}\ \text{kg}\cdot\text{m}^2.$$

Initially, the drive is operating in steady state at its rated speed, supplying its rated torque of 3.2 Nm to the mechanical load connected to its shaft.

At time $t = 0.1$ second, a load-torque disturbance occurs, which causes it to suddenly decrease by 50% (there is no change in load inertia). The feedback control objective is to keep the shaft speed at its initial steady-state value subsequent to the load–torque disturbance. Design the speed feedback controller with the open-loop crossover frequency of 2500 rad/s and a phase margin of 60 degrees. The open-loop crossover frequency of the internal current feedback loop is ten times higher than that of the speed loop and the phase margin is 60°.

Solution

The simulation block diagram is shown in Fig. 10-4 and the Simulink file EX9_1.mdl is included in the accompanying website to this textbook. The simulation results are shown in Fig. 10-5.

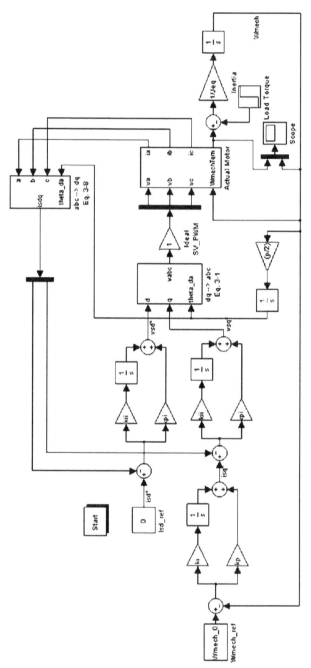

Fig. 10-4 Simulation of Example 10-1.

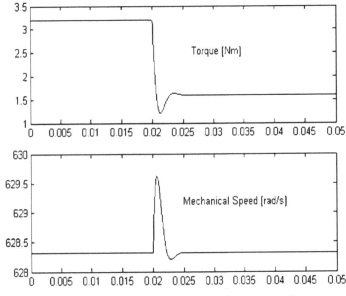

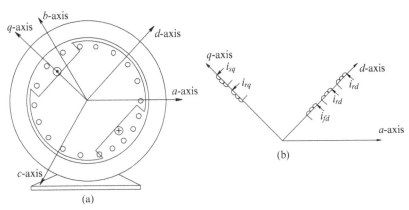

Fig. 10-5   Simulation results of Example 10-1.

Fig. 10-6   Salient-pole machine.

10-3   SALIENT-POLE SYNCHRONOUS MACHINES

Synchronous machines with interior permanent magnets result in unequal reluctance along the d- and the q-axis. In this section, we will go a step further and assume a salient-pole rotor structure as shown in Fig. 10-6a with a rotor field excitation and nonidentical damper windings along the d- and the q-axis. In such a machine, we have the inductances described in the next section.

10-3-1 Inductances

In the stator dq windings,

$$L_{sd} = L_{md} + L_{\ell s} \tag{10-18}$$

and

$$L_{sq} = L_{mq} + L_{\ell s}, \tag{10-19}$$

where the magnetizing inductance of the d winding is not equal to that of the q winding ($L_{md} \neq L_{mq}$) due to the nonsalient nature of the rotor. However, both windings have the same leakage inductance $L_{\ell s}$, which is not affected by the rotor structure.

As shown in Fig. 10-6b, we will replace the actual field winding of N_f turns in the rotor by an equivalent field winding with $\sqrt{3/2}N_s$ turns (where N_s equals the number of turns in each phase of the stator windings), supplied by an equivalent field-winding current i_{fd}. This procedure results in the equivalent field winding having the same magnetizing inductance L_{md} as the stator d winding, hence the equivalent field winding inductance can be written as

$$L_{fd} = L_{md} + L_{\ell fd}, \tag{10-20}$$

where $L_{\ell fd}$ is the leakage inductance of the equivalent field winding.

Similarly in Fig. 10-6b, replacing the actual damper windings with equivalent damper windings, each with $\sqrt{3/2}N_s$ turns, we can write the following equations for the inductances of the equivalent rotor damper windings (with a subscript "r"):

$$L_{rd} = L_{md} + L_{\ell rd} \tag{10-21}$$

and

$$L_{rq} = L_{mq} + L_{\ell rq}, \tag{10-22}$$

where $L_{\ell rd}$ and $L_{\ell rq}$ are the leakage inductance of the equivalent rotor damper windings.

10-3-2   Flux Linkages

In terms of these inductances, the flux linkages of the various windings can be expressed as follows:

Stator *dq* Winding Flux Linkages

$$\lambda_{sd} = L_{sd}i_{sd} + L_{md}i_{rd} + L_{md}i_{fd} \qquad (10\text{-}23)$$

and

$$\lambda_{sq} = L_{sq}i_{sq} + L_{mq}i_{rq}. \qquad (10\text{-}24)$$

Rotor *dq* Winding Flux Linkages

$$\lambda_{rd} = L_{rd}i_{rd} + L_{md}i_{sd} + L_{md}i_{fd} \quad (d\text{-axis damper}) \qquad (10\text{-}25)$$

$$\lambda_{rq} = L_{rq}i_{sq} + L_{mq}i_{sq} \quad (q\text{-axis damper}) \qquad (10\text{-}26)$$

and

$$\lambda_{fd} = L_{fd}i_d + L_{md}i_{sd} + L_{md}i_{rd}. \qquad (10\text{-}27)$$

10-3-3   Winding Voltages

In terms of the above flux linkages, winding voltages can be written as follows, assuming that $\omega_d = \omega_m$ in order to keep the *d*-axis aligned with the rotor magnetic axis.

Stator *dq* Winding Voltages

$$v_{sd} = R_s i_{sd} + \frac{d}{dt}\lambda_{sd} - \omega_m\lambda_{sq} \qquad (10\text{-}28)$$

and

$$v_{sq} = R_s i_{sq} + \frac{d}{dt}\lambda_{sq} + \omega_m\lambda_{sd}. \qquad (10\text{-}29)$$

Rotor *dq* Winding Voltages

$$\underbrace{v_{rd}}_{(=0)} = R_{rd}i_{rd} + \frac{d}{dt}\lambda_{rd} \qquad (10\text{-}30)$$

$$\underbrace{v_{rq}}_{(=0)} = R_{rq}i_{rq} + \frac{d}{dt}\lambda_{rq} \qquad (10\text{-}31)$$

and

$$v_{fd} = R_{fd}i_{fd} + \frac{d}{dt}\lambda_{fd}. \tag{10-32}$$

10-3-4   Electromagnetic Torque

Substituting for flux linkages from above in Eq. (10-6):

$$T_{em} = \frac{p}{2}[\underbrace{L_{md}(i_{fd}+i_{rd})i_{sq}}_{\text{field}+\text{damper in }d\text{-axis}} + \underbrace{(L_{sd}-L_{sq})i_{sd}i_{sq}}_{\text{saliency}} - L_{mq}i_{rq}i_{sd}]. \tag{10-33}$$

10-3-5   *dq*-Axis Equivalent Circuits

Following the procedure used in Chapter 3 for deriving the *dq*-axis equivalent circuits for induction machines, we can draw the equivalent circuits shown in Fig. 10-7 for the *d*- and *q*-axis windings, respectively.

10-3-6   Space Vector Diagram in Steady State

In a balanced sinusoidal steady state with $\omega_d = \omega_m$, the damper winding currents in the rotor are zero, as well as all the time derivatives of

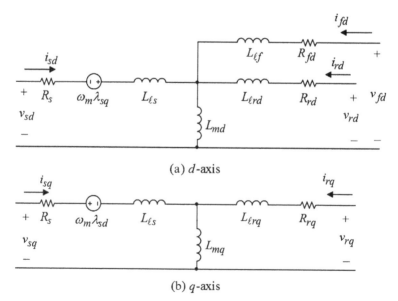

(a) *d*-axis

(b) *q*-axis

Fig. 10-7   Equivalent circuits for a salient-pole machine.

currents and flux linkages in dq windings. Therefore, in Eq. (10-23) and Eq. (10-24),

$$\lambda_{sd} = L_{sd}i_{sd} + L_{md}i_{fd} \qquad (10\text{-}34)$$

and

$$\lambda_{sq} = L_{sq}i_{sq}. \qquad (10\text{-}35)$$

From Eq. (10-28) and Eq. (10-29), using Eq. (10-34) and Eq. (10-35)

$$v_{sd} = R_s i_{sd} - \omega_m L_{sq}i_{sq} \qquad (10\text{-}36)$$

and

$$v_{sq} = R_s i_{sq} + \omega_m L_{sd}i_{sd} + \omega_m L_{md}i_{fd}. \qquad (10\text{-}37)$$

Multiplying Eq. (10-37) by (j) and adding to Eq. (10-36),

$$v_{sd} + jv_{sq} = R_s i_{sd} + jR_s i_{sq} + j\omega_m L_{sd}i_{sd} + j\omega_m L_{md}i_{fd} - \omega_m L_{sq}i_{sq}, \qquad (10\text{-}38)$$

which is represented by a space vector diagram in Fig. 10-8a, where

$$v_{sd} + jv_{sq} = \sqrt{\frac{2}{3}}\vec{v}_s \qquad (10\text{-}39)$$

and

$$i_{sd} + ji_{sq} = \sqrt{\frac{2}{3}}\vec{i}_s. \qquad (10\text{-}40)$$

The corresponding phasor diagram is shown in Fig. 10-8b.

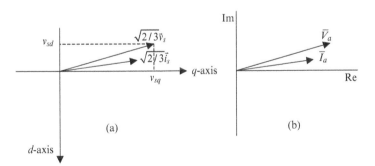

Fig. 10-8   Space vector and phasor diagrams.

10-4 SUMMARY

In this chapter, we have extended the dq-analysis of induction machines to analyze and control synchronous machines.

REFERENCES

1. N. Mohan, *Electric Machines and Drives: A First Course*, Wiley, Hoboken, NJ, 2011. http://www.wiley.com/college/mohan.
2. E.W. Kimbark, *Power System Stability: Synchronous Machines*, IEEE Press, New York, 1995.
3. http://www.baldor.com.

PROBLEMS

10-1 In the simulation of Example 10-1, replace the ideal inverter by an appropriate SV-PWM inverter, similar to that described in Chapter 8.

10-2 Implement flux-weakening in Example 10-1 for extended speed operation.

10-3 Derive the torque expression in Eq. (10-33) for a salient-pole synchronous motor.

11 Switched-Reluctance Motor (SRM) Drives

11-1 INTRODUCTION

In the previous course [1], we have studied variable reluctance stepper motors, whose construction requires a salient stator and a salient rotor. Stepper motors are generally used for position control in an open-loop manner, where by counting the number of electrical pulses supplied and knowing the step angle of the motor, it is possible to rotate the shaft by a desired angle without any feedback. In contrast, switched-reluctance motor (SRM) drives, also doubly salient in construction, are intended to provide continuous rotation and compete with induction motor and brushless dc motor drives in certain applications, such as washing machines and automobiles, with many more applications being contemplated.

In this chapter, we will briefly look at the basic principles of SRM operation and how it is possible to control them in an encoderless manner.

11-2 SWITCHED-RELUCTANCE MOTOR

Cross-section of a four-phase SRM is shown in Fig. 11-1, which looks identical to a variable-reluctance stepper motor. It has a four-phase winding on the stator. In order to achieve a continuous rotation, each phase winding is excited by an appropriate current at an appropriate rotor angle, as well as de-excited at a proper angle. For rotating it in the counterclockwise direction, the excitation sequence is *a-b-c-d*.

Advanced Electric Drives: Analysis, Control, and Modeling Using MATLAB/Simulink®, First Edition. Ned Mohan.
© 2014 John Wiley & Sons, Inc. Published 2014 by John Wiley & Sons, Inc.

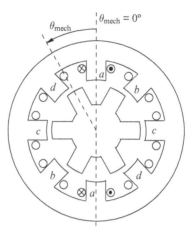

Fig. 11-1 Cross-section of a four-phase 8/6 switched reluctance machine.

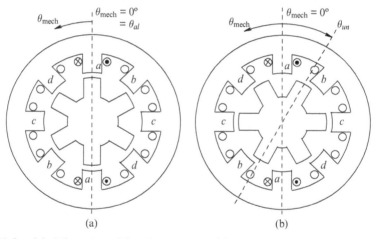

(a)   (b)

Fig. 11-2 (a) Aligned position for phase a; (b) unaligned position for phase a.

An SRM must be designed to operate with magnetic saturation and the reason to do so will be discussed later on in this chapter. Fig. 11-2 shows the aligned and the unaligned rotor positions for phase a. For phase a, the flux linkage λ_a as a function of phase current i_a is plotted in Fig. 11-3 for various values of the rotor position. In the unaligned position where the rotor pole is midway between two stator poles (see Fig. 11-2b, where θ_{mech} equals θ_{un}), the flux path includes a large air gap, thus the reluctance is high. Low flux density keeps the magnetic structure in its linear region, and the phase inductance has a small value.

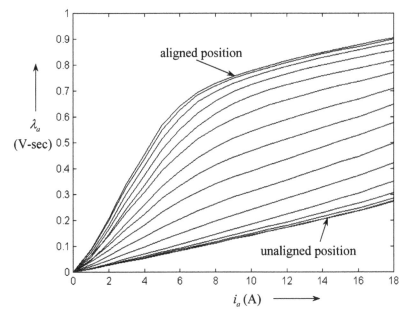

Fig. 11-3   Typical flux linkage characteristics of an SRM.

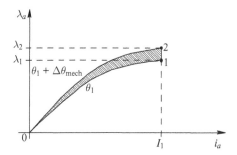

Fig. 11-4   Calculation of torque.

As the rotor moves toward the aligned position of Fig. 11-2a (where θ_{mech} equals zero), the characteristics become progressively more saturated at higher current values.

11-2-1   Electromagnetic Torque T_{em}

With the current built up to a value I_1, as shown in Fig. 11-4, holding the rotor at a position θ_1 between the unaligned and the aligned positions, the instantaneous electromagnetic torque can be calculated as follows: Allowing the rotor to move incrementally under the influence of the electromagnetic torque from position θ_1 to $\theta_1 + \Delta\theta_{mech}$, keeping the current constant at I_1, the incremental mechanical work done is

$$\Delta W_{mech} = T_{em}\Delta\theta_{mech}. \tag{11-1}$$

The increment of energy supplied by the electrical source is

$$\Delta W_{elec} = \text{area}\,(1 - \lambda_1 - \lambda_2 - 2 - 1), \tag{11-2}$$

and the incremental increase in energy storage associated with the phase-a winding is

$$\Delta W_{storage} = \text{area}\,(0 - 2 - \lambda_2 - 0) - \text{area}\,(0 - 1 - \lambda_1 - 0). \tag{11-3}$$

The mechanical work performed is the difference of the energy supplied by the electrical source minus the increase in energy storage

$$\Delta W_{mech} = \Delta W_{elec} - \Delta W_{storage}. \tag{11-4}$$

Therefore in Eq. (11-4),

$$
\begin{aligned}
T_{em}\Delta\theta ={}& \text{area}\,(1 - \lambda_1 - \lambda_2 - 2 - 1) - \{\text{area}\,(0 - 2 - \lambda_2 - 0) \\
& - \text{area}\,(0 - 1 - \lambda_1 - 0)\} \\
={}& \underbrace{\{\text{area}\,(1 - \lambda_1 - \lambda_2 - 2 - 1) + \text{area}\,(0 - 1 - \lambda_1 - 0)\}}_{\text{area}\,(0 - 1 - 2 - \lambda_2 - 0)} \quad (11\text{-}5) \\
& - \text{area}\,(0 - 2 - \lambda_2 - 0) \\
={}& \text{area}\,(0 - 1 - 2 - 0),
\end{aligned}
$$

which is shown shaded in Fig. 11-4. Therefore,

$$T_{em} = \frac{\text{area}\,(0 - 1 - 2 - 0)}{\Delta\theta_{mech}}, \tag{11-6}$$

which is in the direction to increase this area. The area between the $\lambda - i$ characteristic and the horizontal current axis is usually defined as the co-energy W'. Therefore, the area in Eq. (11-6) shown shaded in Fig. 11-4 represents an increase in co-energy. Thus, on a differential basis, we can express the instantaneous electromagnetic torque developed by this motor as the partial derivative of co-energy with respect to the rotor angle, keeping the current constant

$$T_{em} = \left.\frac{\partial W'}{\partial \theta_{mech}}\right|_{i_a=\text{constant}}. \tag{11-7}$$

11-2-2 Induced Back-EMF e_a

With phase a excited by i_a, the movement of the rotor results in a back-emf e_a, and the voltage across the phase-a terminals includes the voltage drop across the resistance of the phase winding:

$$v_a = Ri_a + e_a \tag{11-8}$$

and

$$e_a = \frac{d}{dt}\lambda_a(i_a, \theta_{mech}), \tag{11-9}$$

where the phase winding flux linkage is a function of the phase current and the rotor position, as shown in Fig. 11-3. In terms of partial derivatives, we can rewrite the back-emf in Eq. (11-9) as:

$$e_a = \left.\frac{\partial \lambda_a}{\partial i_a}\right|_{\theta_{mech}} \frac{d}{dt}i_a + \left.\frac{\partial \lambda_a}{\partial \theta_{mech}}\right|_{i_a} \frac{d}{dt}\theta_{mech}, \tag{11-10}$$

where it is important to recognize that a partial derivative with respect to one variable is obtained by keeping the other variable constant.

In Fig. 11-3, the movement of the rotor by an angle $\Delta\theta_{mech}$, keeping the current constant results in a back-emf, which from Eq. (11-10) can be written as:

$$e_a = \left.\frac{\partial \lambda_a}{\partial \theta_{mech}}\right|_{i_a} \underbrace{\frac{d}{dt}\theta_{mech}}_{\omega_{mech}} = \left.\frac{\partial \lambda_a}{\partial \theta_{mech}}\right|_{i_a} \omega_{mech} \left(\frac{d}{dt}i_a = 0\right), \tag{11-11}$$

where ω_{mech} is the instantaneous rotor speed. However, the instantaneous power $(e_a i_a)$ is not equal to the instantaneous mechanical output due to the change in stored energy in the phase winding.

11-3 INSTANTANEOUS WAVEFORMS

For clear understanding, we will initially assume an idealized condition where it is possible to supply the phase winding with a current i_a that has a rectangular waveform as a function of θ_{mech}, as shown in Fig. 11-5. The current is assumed to be built up instantaneously (this will require infinite voltage) at the unaligned position θ_{un} and instantaneously goes to zero at the aligned position θ_{al}. The corresponding waveforms for the electromagnetic torque $T_{em,a}$ and the induced back-emf e_a are also plotted by means of Eq. (11-7) and Eq. (11-11) respectively, with the current held constant.

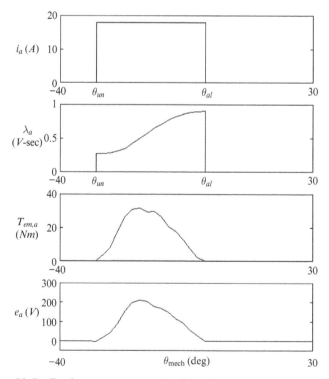

Fig. 11-5 Performance assuming idealized current waveform.

The objectives in selecting these two rotor positions θ_{un} and θ_{al} for current flow are twofold: (1) to maximize the average torque per ampere, and (2) to build up the current to its desired level while the back-emf is small. We can appreciate that with the current flow prior to the unaligned position and after the aligned position, the instantaneous torque would be negative, which would be counter to our objective of maximizing the average torque per ampere. At the unaligned position, the winding inductance is the lowest, and it is easier to build up current in that position compared with other rotor positions.

To achieve instantaneous build-up and decay of phase current assumed in the plots of Fig. 11-5 would require that an infinite phase voltage (positive and negative) is available. In reality with a finite voltage available from the power processing unit to the motor, the phase current waveform for a four-phase motor may look as shown in Fig. 11-6, with the corresponding flux linkage and torque waveforms.

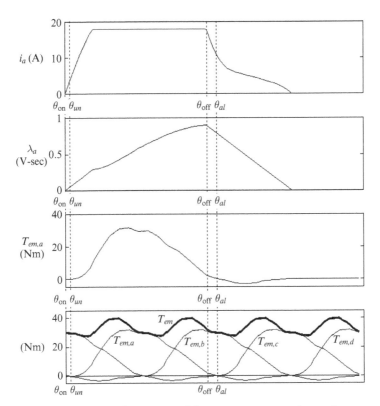

Fig. 11-6 Performance with a power-processing unit.

The phase current build-up is started at an angle θ_{on} (prior to the unaligned position), and the current decay is started at an angle θ_{off} (prior to the aligned position). In order to reduce torque ripple, there is generally overlapping of phases where during a short duration, two of the phases are simultaneously excited. The bottom part in Fig. 11-6 shows the resultant electromagnetic torque T_{em} by summing the torque developed by each of the four phases.

11-4   ROLE OF MAGNETIC SATURATION [2]

Magnetic saturation plays an important role in SRM drives. During each excitation cycle, a large ratio of the energy supplied to a phase winding should be converted into mechanical work, rather than returned to the electrical source at the end of the cycle. We will call this an energy conversion factor. In Fig. 11-7, assuming magnetic saturation and a finite voltage available from the power-processing unit, this factor is as follows:

$$\text{Energy Conversion Factor} = \frac{W_{em}}{W_{em} + W_f}. \tag{11-12}$$

As can be seen from Fig. 11-7, this factor would be clearly higher in the idealized case where an instantaneous build-up and decay of

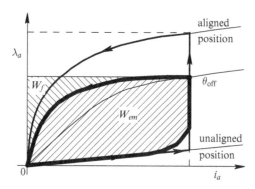

Fig. 11-7   Flux linkage trajectory during motoring.

phase current is assumed. However, without saturation this energy conversion factor is limited to a value of nearly 50%. Magnetic saturation also keeps the rating of the power processor unit from becoming unacceptable.

It should be noted that the Energy Conversion factor is *not* the same as energy efficiency of the motor, although there is a correlation—a lower energy conversion factor means that a larger fraction of energy sloshes back and forth between the power-processing unit and the machine, resulting in power losses in the form of heat and thus in a lower energy efficiency.

11-5 POWER PROCESSING UNITS FOR SRM DRIVES

A large number of topologies for SRM converters have been proposed in the literature. Fig. 11-8 shows a topology that is most versatile. For current build-up, both transistors are turned on simultaneously. (This also shows the robustness of the SRM drive power processing unit, where turning on both transistors simultaneously is normal, which can be catastrophic in other drives.) To maintain the current within a hysteretic band around the reference value, either one of the transistors is turned off, thus making the current freewheel through the opposite diode, or both transistors are tuned off, in which case the current flows into the dc bus and decreases in magnitude. The later condition is also used to quickly de-energize a phase winding.

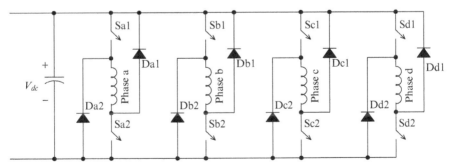

Fig. 11-8 Power converter for a four-phase switched reluctance drive.

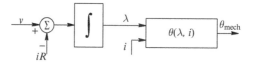

Fig. 11-9 Estimation of rotor position.

11-6 DETERMINING THE ROTOR POSITION FOR ENCODERLESS OPERATION

It is necessary to determine the rotor position so that the current build-up and decay can be started at rotor positions θ_{on} and θ_{off}, respectively, for each phase. There are various methods proposed in the literature to determine the rotor position. One of the easiest methods to explain is shown by means of Fig. 11-3 and Fig. 11-9. The motor can be characterized to achieve the family of curves shown in Fig. 11-3, where the flux linkage of a phase is plotted as a function of the phase current for various values of the rotor position. From this, information, knowing the flux linkage and the current allows the determination of the rotor position. In Fig. 11-9, the flux linkage of an excited phase is computed by integrating the difference of the applied phase voltage and the voltage drop across the winding resistance (see Eq. 11-8). The combination of the measured phase current and the estimated flux linkage then determines the rotor position, using the information of Fig. 11-3.

11-7 CONTROL IN MOTORING MODE

A simple block diagram for speed control is shown in Fig. 11-10, where the actual rotor position is either sensed or estimated using the method described in the previous section or some other technique. The speed error between the reference speed and the actual speed is amplified by means of a PI (proportional-integral) controller to generate a current reference. The rotor angle determines which phases are to be excited, and their current is controlled to equal the reference current as much

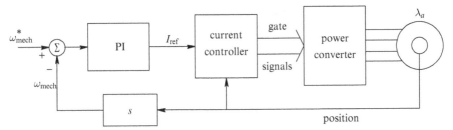

Fig. 11-10   Control block diagram for motoring.

as possible in view of the limited dc-bus voltage of the power-processing unit shown in Fig. 11-8.

11-8 SUMMARY

This chapter discusses SRM drives, doubly salient in construction, which are intended to provide continuous rotation and compete with induction motor and brushless dc motor drives in certain applications, such as washing machines and automobiles, with many more applications being contemplated. In this chapter, we briefly looked at the basic principles of SRM operation and how it is possible to control them in an encoderless manner.

REFERENCES

1. N. Mohan, *Electric Machines and Drives: A First Course*, Wiley, Hoboken, NJ, 2011. http://www.wiley.com/college/mohan.
2. T.J.E. Miller, ed., *Electronic Control of Switched Reluctance Machines*, Newnes, Oxford, 2001.

PROBLEMS

11-1 Show that without magnetic saturation, the energy conversion factor in Fig. 11-7 would be limited to 50%.

11-2 What would the plot of the phase inductance be as a function of the rotor angle (between the unaligned and the aligned rotor positions) for various values of the phase current.

11-3 Although only the motoring mode is discussed in this chapter, SRM drives (like all other drives) can also be operated in a generator mode. Explain how this mode of operation is possible in SRM drives.

INDEX

Note: Page numbers in *italics* refer to figures.

Advanced Electric Drives: Analysis, Control, and Modeling Using
MATLAB/Simulink®, First Edition. Ned Mohan.
© 2014 John Wiley & Sons, Inc. Published 2014 by John Wiley & Sons, Inc.